Contents

Introduction

This book provides a detailed and up-to-date review of rivers and river management for students of AS/A-level geography. It is primarily aimed at sixth-form students, but the content of the book may also be useful for first year undergraduates.

Rivers are the most important natural agents shaping the landscape today. That is why rivers and river management are key topics at AS and A-level. In some specifications they are compulsory; in others they appear as specialist options.

I have tried in this book to stress the dynamic nature of river systems and their responses to changes in energy input and sediment supply. But quite apart from any academic justification for studying rivers, they have great practical importance for millions of people. On the one hand they are resources. They create floodplains and deltas with productive alluvial soils; they provide transport corridors; and they supply water, power and recreational opportunities. On the other hand rivers become hazards when they threaten human activity, most often through flooding and accelerated erosion.

If river-related hazards are to be managed successfully, and rivers used sustainably, an understanding of fluvial processes and systems is crucial. This book is therefore divided into three parts. The first deals with fluvial processes, the second with geomorphology and the third with river management. As we will see, many of the failures of river management in the past stem from inadequate understanding of fluvial processes and systems. Recent approaches to management recognise the need to work with rivers rather than against them. As with coastal management, this new approach puts the emphasis on 'soft' rather then 'hard' structures. In some cases, it has even led to programmes (often controversial) of restoring over-engineered rivers to something approaching their natural state.

As a student, you can use this book as a learning tool in several different ways. Most obviously the book provides knowledge and understanding of how rivers operate and how people interact with them. The text should be read to consolidate your understanding simultaneously with coverage of each topic in class. You will, of course, need to refer to specific areas of the text to complete essays and other assignments.

Don't ignore the case studies. They are particularly important to support general discussion and explanation. They should be used to illustrate not only your assignments, but also your written work in the final examinations. Information is backed up with photos, tables and diagrams. These should be scrutinised just as carefully as the text.

An integral feature of this book is its many activities, through which it becomes interactive. Some activities aim to test your knowledge and understanding; some develop essential skills such as statistical analysis and hypothesis testing; and others encourage you to investigate topics further, through research on the internet or in your school library. And for those who really engage with rivers, there are many ideas that could form the basis for fieldwork investigations and coursework.

Michael Raw

1 Rivers as energy systems

Rivers and streams are natural bodies of water that flow in open channels. They have three important roles in landscape development:

- They erode the channels in which they flow.
- They transport sediments and solutes provided by weathering and slope processes.
- They create distinctive erosional and depositional landforms.

Today, rivers are the dominant natural agents of landscape change in most environments.

Drainage basin systems

Drainage basins or catchments are simple open systems with inputs and outputs of energy (e.g. solar and terrestrial radiation) and matter (e.g. water and sediment). In any catchment, water and sediments move under the effects of gravity towards the lowest point in the basin.

Stream energy drives the drainage basin system. The amount of energy available to a stream depends on:

- the vertical distance to sea level at any point in a stream's course (**potential energy**)
- the volume and velocity of stream flow

As streams flow, they convert potential energy to **kinetic energy**. Normally, 95% of this kinetic energy is spent overcoming the internal friction of water and the frictional drag of the channel bed, banks and coarse sediments (Figure 1.1). The overall amount of energy available to a stream is always equalled by energy expenditure. However, this energy expenditure occurs in a specific order. As Figure 1.1 shows, energy is first spent on **flow**, then on **sediment transport**, and finally on **erosion**.

Table 1.1 shows the power of four streams in northern England at average and high discharge levels. During periods of high discharge, streams have more than enough energy to overcome frictional resistance to flow.

Figure
1.1 **The fluvial energy system**

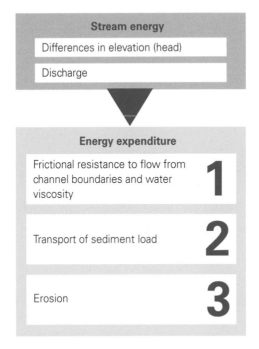

Table
1.1 **Power of selected rivers at average and high discharge levels**

Stream	Station	Vertical drop (m)*	Mean discharge (cumecs)	Power (kW)	High flow discharge (cumecs)	Power (kW)
Wharfe	Addingham	624.3	14.07	69 511	100	493 802
Langdon Beck	Langdon Beck	327.0	0.42	1 087	4	10 350
Kent	Sedgwick	798.1	9.94	56 463	50	315 788
Bela	Beetham	327.1	3.54	9 163	20	51 770

*Difference in height between station and highest point in catchment

Changing energy inputs

Significant increases in energy input lead to corresponding increases in sediment transport and erosion. These increases occur on short, medium and long timescales.

THE HENLEY COLLEGE LIBRARY

Short-term and medium-term changes

Stream energy levels can vary abruptly over short time scales. A violent thunderstorm often produces rapid increases in discharge (Figure 1.2) and stream energy. This has important consequences for sediment transport and erosion. For example, a tenfold increase in discharge raises the transporting power of a stream not 10 times, but 10^3 or 10^4 times!

Figure 1.2 **Short-term increase in stream discharge, Langdon Beck, July 1983**

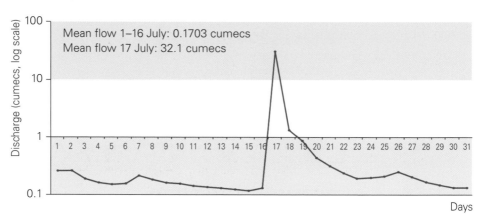

More prolonged rainfall events raise stream power gradually and may sustain high energy levels for several days (Figure 1.3).

Figure 1.3 **Medium-term increase in stream discharge, River Derwent at Buttercrambe, March 1999**

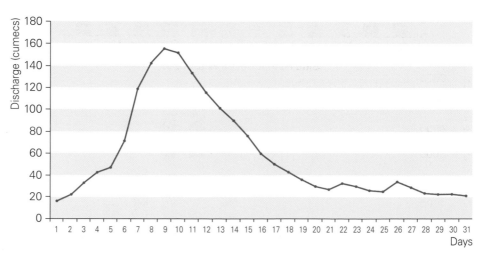

Streams draining glaciated uplands such as the Alps and the Rockies, swollen by meltwater in spring and early summer, have exceptionally high energy levels lasting several weeks. Meanwhile, in monsoonal Asia and tropical continental Africa where rainfall is seasonal, river energy may be sustained at high levels for three or four months (Figure 1.4).

Figure
14

Annual hydrographs for three rivers

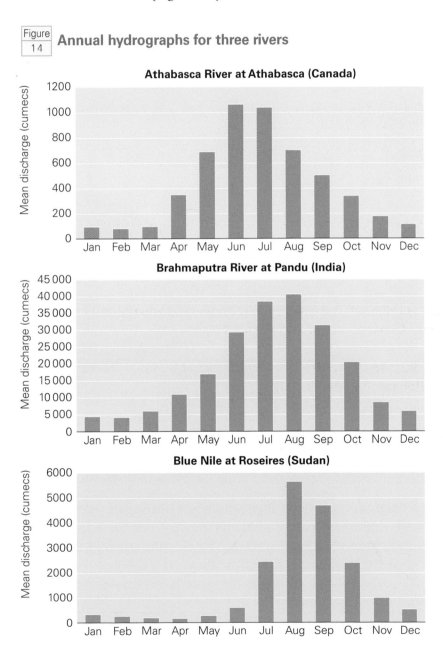

Long-term changes

Fifteen thousand years ago, the climate of Eurasia and North America began to warm as the Earth emerged from the long Devensian glacial. Rivers in highland Britain, swollen by meltwater, had abnormally high discharges and energy levels. As a result, they were able to transport huge volumes of sediment left by receding ice sheets and glaciers. These rivers probably resembled those fed by glacial meltwater in Iceland and the South Island of New Zealand today, with multi-thread channels, shifting gravel bars, and high-energy floodplains (Figure 1.5).

| Figure 1.5 | **High-energy floodplain with braided river fed by meltwater, Thorsmark, Iceland** |

The effect of high discharge and high-sediment loads was to in-fill former glacial valleys with thick layers of sand and gravel. Today, in many parts of northern Britain, rivers are actively eroding these deposits, releasing them from long-term store. For example, on the River Skirfare in North Yorkshire, lateral erosion on the outer bank of a meander has exposed sand and gravel deposits (Figure 1.6).

Figure 1.6 **Sediment storage in river terraces on the River Skirfare, North Yorkshire**

River terrace (ancient floodplain) — sediment store

Coarse gravels derived from post-glacial point bar and channel bar deposits being eroded on the outer bank of a meander

Equilibrium and grade

Rivers and streams are said to reach **dynamic equilibrium** when they have just sufficient energy to transport their sediment load. The idea of equilibrium assumes that the energy expenditure of rivers and streams adjusts to achieve an overall state of balance (Figure 1.7). For example, if a stream has surplus energy it erodes its channel. This reduces its gradient and ultimately its energy levels, until balance is restored. Thus streams, in common with other natural systems, have the ability to self-adjust. We refer to the automatic adjustments that restore stability as **negative feedback**.

Figure 1.7 **Factors affecting channel equilibrium**

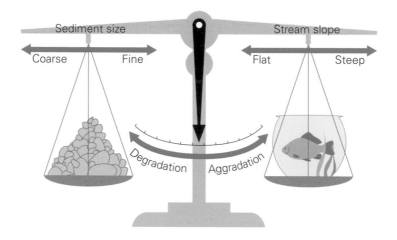

The most striking expression of dynamic equilibrium is a river's **graded long profile** from its source to **base level**. Base level may be sea level, the confluence of a tributary stream with a larger river or, occasionally, a resistant rock bar in the river's course which forms the local base level for the catchment upstream of the bar. For example, a resistant band of limestone interrupts the River Ure's long profile at Aysgarth Falls, North Yorkshire (Figure 1.8).

Figure 1.8 **Aysgarth Falls, River Ure**

Graded long profiles are concave in section and are linked to the downstream increase in energy due to higher discharge. The effect of higher discharge is to increase erosion and eventually lower the channel gradient towards base level. Thus erosion reduces the channel gradient and thereby maintains the balance between stream energy and sediment load.

Activity 1

Study Figuro 1.7. Using evidence from Figure 1 7 only, describe and explain how a stream is likely to respond to:

- an increase in stream slope
- a decrease in stream slope
- a reduction in discharge
- an increase in the coarseness of load
- a decrease in load

Stream responses to changing fluvial environments

Dynamic equilibrium occurs when sediment load (Q_s), sediment particle size (D_{50}), streamflow or discharge (Q_w) and channel gradient or stream slope (S) are in balance. This is summarised by the formula:

$$Q_s\, D_{50} \propto Q_w nS$$

This formula tells us that any change in sediment load and bed material size will be balanced by changes in streamflow or in channel gradient. Hence, change in any one of these variables causes change in one or more of the other variables, until dynamic equilibrium is established. For instance, if sediment load increases and discharge remains constant, **aggradation** (i.e. deposition in the channel) must occur, which increases the channel gradient.

Additional relationships have been suggested in response to change in alluvial channels. For example, width (b), depth (d) and meander wavelength (L) are directly proportional to discharge (Q_w); and channel gradient (S) is inversely proportional to discharge (Q_w):

$$Q_w \propto \frac{bLd}{S}$$

Other relationships exist between sediment load (Q_s) and sinuosity (P), and the above variables. Width, meander wavelength and channel gradient are directly proportional to sediment load; and depth and sinuosity are inversely proportional to sediment load in alluvial channels.

Table 1.2 shows further examples of the relationships between discharge and channel width, depth, gradient, sinuosity and meander wavelength.

| Table 1.2 | Stream responses to changes in alluvial channels |

Formula	Example
$Q_w^+ \propto b^+, d^+, L^+, S^-$	An increase in discharge (e.g. where a tributary joins the main channel; increased runoff due to urbanisation) will lead to an increase in width, depth and meander wavelength, but a decrease in slope. This is because any increase in discharge will bring a corresponding increase in energy expenditure and erosion. Erosion will widen the channel and increase channel depth, but will decrease channel gradient until a new equilibrium is established.
$Q_w^- \propto b^-, d^-, L^-, S^+$	Any decrease in discharge (e.g. resulting from abstraction of water) will decrease the river's energy. The result — deposition — will reduce channel width and depth, but increase channel slope. This will re-establish equilibrium and minimum energy expenditure.
$Q_s^+ \propto b^+, d^-, L^+, S^+, P^-$	An increased sediment load caused by localised bank erosion will disrupt equilibrium. The channel will widen and become shallower through deposition of excess sediment. However, this deposition will steepen the channel gradient and restore equilibrium.
$Q_s^- \propto b^-, d^+, L^-, S^-, P^+$	A reduction in sediment load (e.g. a stream leaving a lake or reservoir) will increase available energy, deepening the channel and reducing the gradient to establish a new equilibrium. Excess energy will be dissipated by an increase in sinuosity.

Human impact on the fluvial energy system

Human activities have important effects on stream discharge and sediment load. Just as rivers adjust to environmental change, they also respond to changes caused by human activities and eventually a new equilibrium is established.

Land-use changes

The conversion of land from rural to urban use increases runoff and stream discharge. Urbanisation also increases stream sediment loads through the construction of roads, bridges and buildings. Changes in agriculture, such as an increase in arable land at the expense of pasture, also increase sediment inputs to streams and rivers. Meanwhile, rates of lateral erosion increase when rivers are isolated from their floodplains (e.g. by levées) or when bank vegetation is destroyed.

Flow regulation and water abstraction

Streams flowing from reservoirs carry little sediment. They expend minimal energy on sediment transport and therefore have surplus energy for erosion. The outcome is often channel incision and the scouring of beds and banks. This process is known as **clearwater erosion**. Vertical erosion may cause rivers to lose access to their floodplains during annual floods. Paradoxically, this may increase the risk of flooding further downstream. Clearwater erosion may also result in streams eroding laterally, widening their channels, which fill with sediment and create new floodplains.

Heavily reservoired catchments are associated with steep reductions in peak flow and energy levels (Figure 1.9). Water abstraction has the same effect, reducing runoff as a percentage of total precipitation. This also has implications for sediment transport and erosion.

Activity 2

The Rivers Ure and Nidd in North Yorkshire drain Pennine catchments and flow eastwards to the River Ouse. The Nidd has three large reservoirs in its catchment. Angram and Scar House supply water to Bradford; Gouthwaite is used as a compensation reservoir. The River Ure has no reservoirs, and just one natural lake — Semer Water — in its catchment.

Study the data in Figure 1.9 showing discharge and precipitation in the Ure and Nidd catchments. Examine the likely impact of reservoirs on river flow and the fluvial energy system.

Channel straightening

Channel straightening and gravel mining of point bars have the effect of steepening stream channels. Steeper channels increase energy levels in the river and lead to scouring of channels and the development of new floodplains at lower elevations. Bed erosion occurs where meanders are straightened and the channel gradient increases. Headcuts (places where incision of the river bed has occurred) migrate upstream, eroding sediments from otherwise stable stream beds. This increases the sediment load, which causes aggradation downstream and the formation of channel bars, lateral scour and the erosion of river banks.

| Figure 1.9 | Human impact on the Ure and Nidd catchments, 2001 |

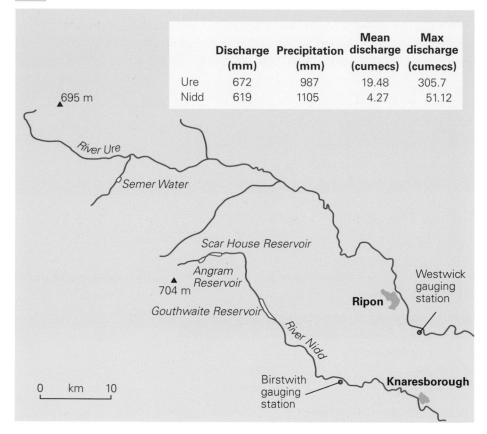

	Discharge (mm)	Precipitation (mm)	Mean discharge (cumecs)	Max discharge (cumecs)
Ure	672	987	19.48	305.7
Nidd	619	1105	4.27	51.12

Construction

It is difficult for streams to achieve a new equilibrium where roads and other infrastructure minimise valley space for the river to access or create a floodplain. Increasingly, landowners and government agencies armour river channels and build levées to contain floodwaters in the channel. Such measures are no more than temporary fixes — sooner or later they lead to problems of accelerated erosion and resource degradation.

2 Runoff and stream flow

THE HENLEY COLLEGE LIBRARY

The water balance

Runoff is the movement of water across the land surface and in channels, as rivers, streams and rivulets. Figure 2.1 shows that precipitation reaches stream channels through four separate pathways:

- directly into stream channels from precipitation
- across the ground surface as **overland flow**
- through the soil as **throughflow**
- through permeable rocks as **groundwater flow**

| Figure 2.1 | **Movement of water in drainage basins**

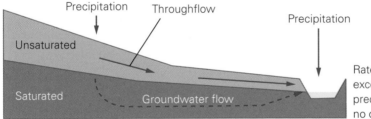

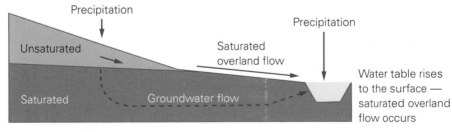

These pathways vary in the speed at which they deliver water to stream channels. Direct channel precipitation is fastest, but accounts for only a tiny fraction of the total stream flow. Overland flow is also fast-moving and occurs either when the soil is saturated or when the intensity of precipitation exceeds the **infiltration capacity** of the soil. Water moving laterally through the soil as **throughflow** takes several hours or days to reach stream channels. Compared with groundwater flow this is rapid — groundwater may be stored for years or even decades before it is released as runoff.

The **water balance** provides a framework for understanding long-term runoff in a drainage basin (Figure 2.2). It states that runoff is equal to precipitation minus evapotranspiration and water entering or leaving storage.

runoff = precipitation – evapotranspiration +/– storage

Figure 2.2 **The River Inver drainage basin**

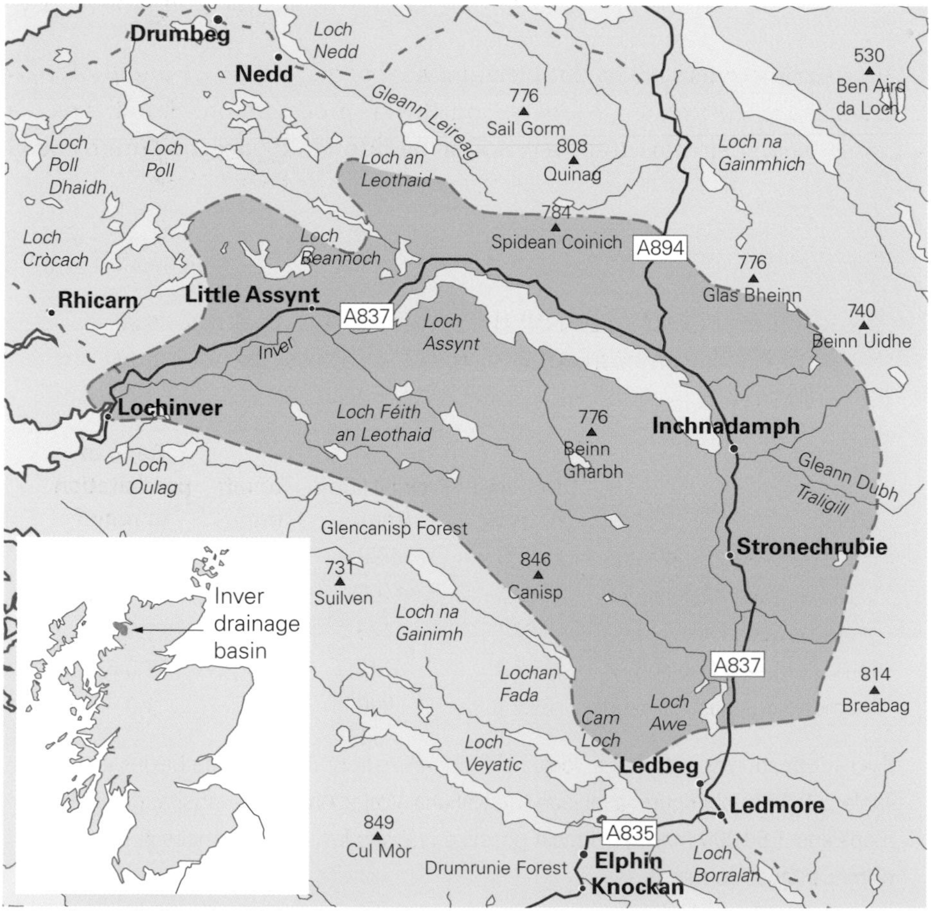

The River Inver in the Assynt area of northwest Scotland (Figure 2.2) is one of the few sizeable catchments in the UK unaffected by human activities. Therefore, it can be used as a simple illustration of the water balance.

The Inver catchment has a mean annual precipitation of 2211 mm, of which 1948 mm eventually flows into the River Inver and its tributaries. This means that just 263 mm are lost to evaporation and transpiration. This relatively small amount is explained by the catchment having:

- a cool, cloudy climate, caused by its high latitude (58°N), a westerly location and mountains rising to over 800 metres
- sparse woodland cover, which reduces interception and transpiration
- steep slopes, which promote rapid runoff
- impermeable rocks, such as gneiss, coarse sandstone and quartzite, which increase rates of overland flow

Average annual discharge figures, such as those in Table 2.1, give us an indication of the power of a river and therefore of its erosional and transportational capacity. However, average annual discharge tells only part of the story.

Short-term events such as thunderstorms, or periods of heavy and prolonged rainfall, have the greatest influence on fluvial processes and landforms. On these occasions, streams and rivers flow at **bankfull** and have maximum energy.

Activity 1

Table 2.1 **Average discharge and the proportion of mean annual precipitation converted to runoff in some UK drainage basins**

River	Discharge (cumecs)	Precipitation (mm)	Runoff (mm)	Ratio of precipitation to runoff
Eden (northwest England)	3.98	799	401	1:0.502
Kent (northwest England)	7.36	1409	1111	1:0.789
Tay (east Scotland)	167.2	1425	1150	1:0.807
Tees (northeast England)	15.58	1141	654	1:0.573
Thames (southeast England)	65.64	806	327	1:0.406

Find information about the geology, relief and climate of the drainage basins in Table 2.1. (Possible sources include the National Water Archive's website, geology maps and 1:50 000 maps.) Suggest possible reasons for the differences in the ratios of precipitation to runoff.

Bankfull discharge

Bankfull discharge is the maximum discharge, measured in cubic metres per second (cumecs), that can be contained by a stream or river channel (Figure 2.3). Any increase in discharge above bankfull results in overtopping of the banks and flooding of the adjacent valley floor. Bankfull flows occur infrequently, usually not more than once or twice a year.

Figure 2.3 **River Aln, Northumberland**

Bankfull discharge

Low flow conditions

Bankfull flow is now recognised as the principal control of channel shape and channel landforms. This is because at bankfull, streams and rivers have maximum energy to erode their channels and transport sediment. Bankfull discharge moves the most sediment and water for the least amount of energy expended. However, bankfull flow is not the same as highest flow. The highest flows are recorded when rivers flood and water moves through valleys along **floodways**. Highest flows are less important geomorphologically than bankfull flows. This is because water spilling from the channel onto the floodplain loses energy to friction. Despite the higher discharge, this loss of energy reduces the erosional and transporting potential of the river and increases the likelihood of deposition.

Measuring bankfull cross-sectional channel shape

The first task when measuring a bankfull channel in cross-section is to identify the bankfull channel in the field (Figure 2.4). Indicators of bankfull water levels include:
- the tops of active depositional surfaces in the channel (e.g. channel bars)
- the level of wash or scour of exposed tree roots
- trash lines
- the height of persistent woody plants such as alders and willows

Figure 2.4   Hydrologic and topographic floodplains

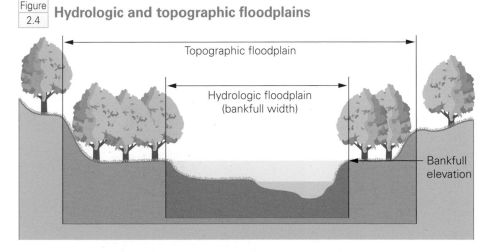

An ideal bankfull channel for measurement has the following characteristics:
- a single (not multiple) channel
- few bank obstructions, such as tree roots and thick vegetation
- opposite banks of equal height
- a roughly straight planform

■ few obstructions in the channel, such as boulders and coarse bedload particles

It is clear that many sites are unsuitable for bankfull measurements. For example, in a meander, the outer (cut) bank is usually at a higher level than the inner bank (point bar). Equally, where a river has incised its channel and abandoned its original floodplain, the distinction between the channel banks and terraces may be unclear.

Activity 2

Study Figures 2.5(a) and 2.5(b) and assess their suitability for bankfull cross-sectional measurement.

Figure 2.5 **(a) Lupton Beck, Cumbria**

(b) Traligill Beck, Sutherland

Energy efficiency and bankfull cross-sectional shape

The shape of a stream channel in cross-section affects its efficiency as a conveyor of water and sediment. The most efficient cross-sectional shape is that which minimises energy loss through friction between flowing water and the channel bed and banks. Frictional resistance is lowest when the ratio between the **wetted perimeter** (i.e. the perimeter of the bankfull channel in cross-section) and the cross-sectional area of the channel is high.

This ratio is the **hydraulic radius** (r) and is the most commonly used measure of channel efficiency. The greater the value of the hydraulic radius, the more efficient the channel is.

$$\text{hydraulic radius } (r) = \frac{\text{bankfull cross-sectional area (m}^2)}{\text{bankfull wetted perimeter (m)}}$$

An alternative, simpler measure of channel efficiency is **width-to-depth ratio**. This is obtained by dividing channel width by average depth. In contrast to the hydraulic radius, an efficient channel has a low width-to-depth ratio value.

Downstream changes in channel efficiency

In general, stream channels become more efficient downstream. This change is an adjustment to increasing discharge downstream. As discharge increases, the cross-sectional area of stream channels increases faster than width and depth. Thus the ratio of channel cross-sectional area to wetted perimeter, and therefore channel efficiency, must increase.

Activity 3

Table 2.2 **Bankfull channel cross-sections, River Aire at Kirkby Malham and at Gargrave, North Yorkshire**

	Width (m)	Depth at equal intervals across the channel (m)								
Kirkby Malham	8.8	0.32	0.46	0.55	0.68	0.83	0.84	0.72	0.68	0.22
Gargrave	18.3	0.70	0.79	1.42	1.99	2.08	2.09	1.85	1.62	0.87

(a) Use the data in Table 2.2 to draw the two bankfull channels on graph paper or in Excel. Use the same scale for width and depth.

(b) Estimate the bankfull cross-sectional areas and the bankfull wetted perimeters, and calculate the hydraulic radius (r) for each channel.

(c) Compare the hydraulic radii of the channels and comment on the relative efficiency of the channels.

Process

Activity 4

This fieldwork investigation aims to:
- measure cross-sectional channel shape and efficiency
- compare upstream and downstream changes in cross-sectional channel shape and efficiency

(a) Select a 100 m upstream reach of river and a 100 m downstream reach, making sure that the two reaches are comparable in terms of bank material, planform and bank vegetation, and free from human influence. (NB Make sure you complete a full risk assessment of each site.)

(b) Measure ten channel cross-sections chosen (randomly or systematically) along each reach. (Bankfull depth should be measured at 10% intervals across the channel.) Calculate the hydraulic radius and width-to-depth ratio for each cross-section.

(c) Test for the significance of differences between upstream and downstream channel shapes using the Mann–Whitney U-test.

(d) Evaluate your methodology and results.

Activity 5

| Table 2.3 | Hydraulic radii at upstream and downstream sites, Eglingham Burn, Northumberland |

Upstream site	Downstream site	Upstream site	Downstream site
0.356	0.518	0.440	0.703
0.596	0.667	0.430	0.577
0.353	0.542	0.361	0.737
0.827	0.927	0.404	0.917
0.449	0.818	0.491	0.691
0.459	0.584	0.475	0.779
0.451	0.791	0.320	0.831
0.588	0.725	0.511	0.994
0.474	0.824	0.592	0.839

The data in Table 2.3 can be used to test the hypothesis that the channel efficiency of Eglingham Burn increases downstream.

(a) Plot the data at upstream and downstream sites as a dispersion diagram.

(b) Analyse the statistical significance of the difference between the two data sets using the Mann–Whitney U-test.

(c) Comment on the results.

Estimating bankfull discharge

Discharge is the product of channel cross-sectional area (m²) and flow velocity (m s⁻¹).

Accurate estimates of channel cross-sectional area can be obtained at low flow by careful measurement in the field. However, the calculation of bankfull discharge also requires an estimate of bankfull flow velocity.

Because of the practical difficulties of measuring velocity in high-flow conditions, Manning's equation is used to estimate bankfull velocity:

$$\text{velocity } (v) = \frac{r^{0.66}\, s^{0.5}}{n}$$

where

r = hydraulic radius

s = slope of water surface (tangent of the angle) — the gradient of the adjacent valley floor is normally used as a substitute for the slope of the water surface

n = coefficient of roughness (see Table 2.4)

Table 2.4 **Roughness coefficients for Manning's equation for bankfull velocity**

Channel type	Coefficient range
Small mountain stream, pebble and boulder bed	0.040–0.070
Small, clean, straight lowland stream	0.025–0.033
Small, weedy stream with deep pools	0.075–0.150
Floodplain stream in pasture land	0.025–0.035
Floodplain stream in heavy woodland	0.100–0.150
Large stream (width >33 m)	0.025–0.060

Estimates of bankfull velocity from Manning's equation are based on measurements of the gradient of the valley floor (s) and the resistance of the channel to flow. The latter is represented by the hydraulic radius (r) and channel roughness (n). Thus, streams with steep gradients, near-semicircular channels in cross-section and fine, well-sorted bedload are likely to have high bankfull velocities.

Lower bankfull velocities are a feature of streams with gentle gradients, wide and shallow channels and coarse, poorly sorted bedload.

Activity 6

(a) Estimate the roughness coefficients for the stream channels in Figures 2.3, 2.5 and 2.6.

(b) Using the data for Eglingham Burn in Table 2.5, calculate:
- the bankfull velocity
- the bankfull discharge for upstream and downstream sites

Table 2.5 Channel parameters: upstream and downstream sites on Eglingham Burn, Northumberland

	Upstream site	Downstream site
Average width (m)	5.223	8.575
Average depth (m)	0.565	0.877
Slope (°)	0.5	0.5
Tangent (gradient) of slope	0.009	0.009
Hydraulic radius	0.484	0.756
Coefficient of roughness	0.04	0.03
Estimated bankfull discharge ($m^3 s^{-1}$)	2.904	9.988

Figure 2.6 Coarse bedload in a stream channel, Marshaw Wyre, Lancashire

Hydraulic geometry

Hydraulic geometry describes the changes in stream channel characteristics as bankfull discharge increases.

Downstream changes

Downstream changes in hydraulic geometry are based on spatial variations in channel cross-section characteristics. As distance downstream increases, there are corresponding increases in the size of the drainage basin and therefore in discharge and sediment load. The relationship between changes in channel shape and downstream increases in bankfull flow has been intensively studied. These studies show systematic increases in channel width and channel depth with increasing bankfull discharge. They also show that channel width normally increases more rapidly than channel depth (Figure 2.7).

Although the associations between bankfull discharge and channel width and depth are strong, they are not perfect. This is because there are factors other than discharge that influence changes in cross-sectional channel shape, including: sediment load and size; bank materials; bank vegetation; and human activities.

The influence of sediment load and sediment size on channel shape is less obvious than that of bankfull discharge. However, coarse sediment loads are invariably associated with wide, shallow channels and high width-to-depth ratios. On the other hand, streams with a higher percentage of fine suspended load tend to have relatively narrow, deep channels.

Bank materials also have an important effect on channel width. Easily erodible materials such as sands and gravels tend to produce wider channels than cohesive silts and clays. Channels with alluvial beds are likely to be deeper than those where resistant materials such as solid rock limit vertical erosion. Vegetation (especially trees) stabilises the banks of streams and greatly reduces the rate of lateral erosion. Finally, human activities such as erosion control and bank reinforcement have an obvious, if limited, influence on channel shape.

At-a-station changes

At-a-station changes relate to variations in discharge over time at a specific location and the effects of these changes on the width and depth of water in the channel, and on flow velocities. These changes depend to a large extent on bankfull channel shape.

Hydraulic geometry of Dartmoor streams

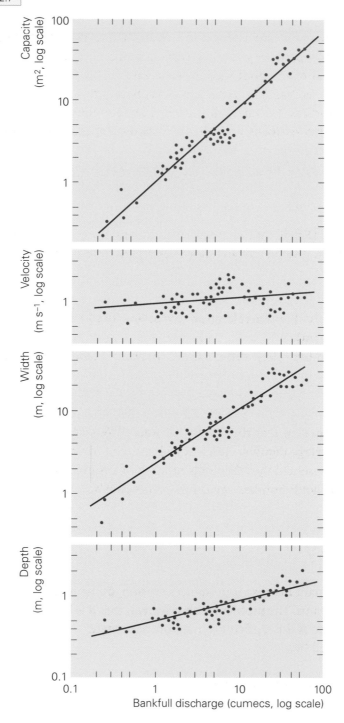

Flow hydraulics

The flow of water in stream channels can be either **laminar** or **turbulent**. When flow is laminar, the water particles glide past each other smoothly and uniformly. Laminar flow is typical of the pool sections of stream and river channels. In contrast, turbulent flow is chaotic — water particles collide and change direction, creating eddies and vortices. Turbulent flow dominates mountain streams, with their boulder-beds and steep slopes, and helps to dissipate surplus energy as heat.

Measuring types of flow: Froude and Reynolds numbers

The **Froude number (Fr)** identifies types of turbulent flow. A value of <1 indicates tranquil flow. When Fr >1 the flow is extremely turbulent.

$$Fr = \frac{v}{\sqrt{gd}}$$

Where v is velocity (m s^{-1}); g is acceleration due to gravity (9.8 m s^{-2}); d is the average depth of flow (m).

The **Reynolds number (Re)** differentiates between laminar flow and turbulent flow.

$$Re = \frac{pvr}{u}$$

Where v is velocity (m s^{-1}); P is the density of water (1000 kg m^{-3}); r is the hydraulic radius or average depth (m); u is the viscosity of water (0.01139).

Laminar flow is represented by a Reynolds number of around 500. However, most streams have Reynolds numbers greater than 2000 and have turbulent flow.

Flow velocity

Flow velocity is important because it influences erosion, transport and depositional processes in streams. The problem with velocity is that it is highly variable both in space and time and is difficult to measure accurately.

Velocity is influenced by:
- channel gradient
- channel shape (in cross-section and planform)
- sediment load

Activity 7

Table 2.6 Flow types, Marshaw Wyre, Lancashire

Site	Average depth (m)	Velocity (m s^{-1})	Acceleration (m s^{-2})	Froude number	Reynolds number
1	0.074	0.357	9.8		
2	0.138	0.447	9.8	0.38	
3	0.142	0.439	9.8		
4	0.163	0.188	9.8		2690.4
5	0.167	0.434	9.8		

Table 2.7 Flow types, Feldon Burn, County Durham

Site	Average depth (m)	Velocity (m s^{-1})	Acceleration (m s^{-2})	Froude number	Reynolds number
1	0.15	0.48	9.8		
2	0.08	0.69	9.8	0.78	4846.3
3	0.05	0.55	9.8		
4	0.10	0.49	9.8		
5	0.19	0.47	9.8		

(a) Complete the calculations of the Froude and Reynolds numbers in Tables 2.6 and 2.7. Comment on your results.

(b) Measure the flow characteristics (velocity, mean depth) of a sequence of pools and riffles on a gravel bed stream and calculate the Froude and Reynolds numbers.

(c) Comment on the differences in flow numbers between pools and riffles as indicated by the Froude and Reynolds numbers.

(d) Analyse the significance of flow difference using the Mann–Whitney U-test.

Velocity is proportional to channel slope and inversely proportional to the frictional resistance of channel boundaries and the volume and calibre of transported sediment. The increase in velocity that occurs with discharge is related to:

- the increase in depth and reduction in the ratio of wetted perimeter to cross-sectional area — overall there is a significant decrease in frictional resistance
- a decrease in sinuosity and therefore an increase in average gradient

One reason why velocity is so hard to measure is because it varies in three dimensions — with depth, across the stream and downstream. Water flowing in channels is retarded by the friction of the bed and banks, so velocities increase away from channel boundaries. Look at the depth profile in Figure 2.8. It shows that velocity is at a minimum close to the stream bed and reaches a maximum at the surface. Velocity also increases towards the centre of a stream as the frictional effect of the channel banks declines. Channel cross-section velocity also depends on the shape and alignment of the channel. In straight reaches, the velocity pattern is symmetrical; in meanders it is characteristically asymmetric.

Figure 2.8 **Velocity cross-profiles showing isovels (m s^{-1}) in (a) straight and (b) meandering channels**

(a)
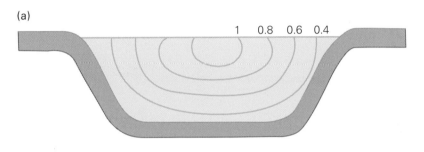

(b)
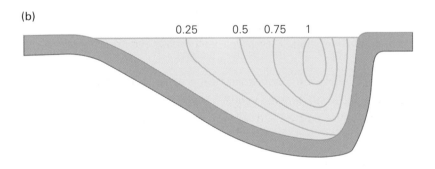

Although the gradient of streams decreases downstream, velocity tends either to remain constant or to increase slightly. This is because the channel becomes hydraulically more efficient downstream, thus reducing resistance to flow.

Activity 8

Figure 2.9 **Downstream changes in discharge, width, depth, velocity and gradient**

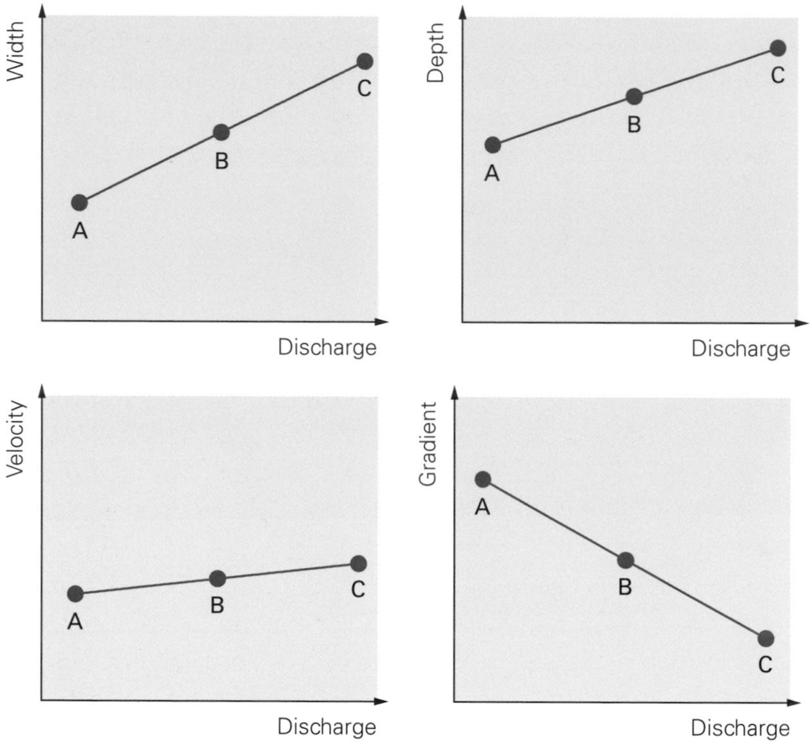

A, B and C are located sequentially downstream, where
- A = sediment supply zone (headwaters)
- B = sediment transfer zone
- C = sediment storage zone

Study Figure 2.9, which summarises the downstream relationships between discharge and channel width, depth, gradient and velocity.

(a) Describe and explain the relationship shown in each graph in Figure 2.9.

(b) Explain why velocity increases downstream, despite the decrease in gradient.

3 Fluvial processes

In Chapter 1 we saw that the kinetic energy of streams is primarily spent on overcoming the frictional resistance to flow of water particles and boundary materials. Only then, if surplus energy is available, will sediment transport and channel erosion occur.

Erosion

Fluvial erosion is the removal of rock and other mineral particles from the channel bed and banks by running water. As velocity increases, a condition is reached in which the forces tending to move particles are exactly balanced by the resisting forces (friction). This condition is called the **critical erosion threshold**. Any increase in velocity beyond this point results in erosion.

In addition to the downstream force exerted by moving water, water flowing near the stream bed also creates a lifting force. This force results from differences in flow velocity between the top and bottom of a particle. It creates a pressure gradient that causes particles to move vertically. In turbulent flow conditions, the lifting force is also aided by localised eddying, which acts directly upwards from the bed.

The erosion and transport of sediments depends on river energy (or flow velocity), turbulence and sediment size. The Hjulström curve (Figure 3.1) shows the velocities needed for the **entrainment** and movement of particles. Generally, the higher the velocity the larger are the particles transported. However, tiny silt and clay particles with diameters of less than 0.05 mm are an exception.

The upper curve in Figure 3.1 shows that the smallest silt and clay particles have higher erosion velocities than larger sand-sized particles. For example, medium-sized particles such as sand are eroded at velocities of just 0.2–0.3 m s^{-1}, while clay particles only erode when velocities reach 1–2 m s^{-1}. The explanation of this anomaly is the cohesiveness of clays — they stick together because of electrical bonding. Gravels, which form part of the bedload, need high velocities of 3 m s^{-1} and above for transport.

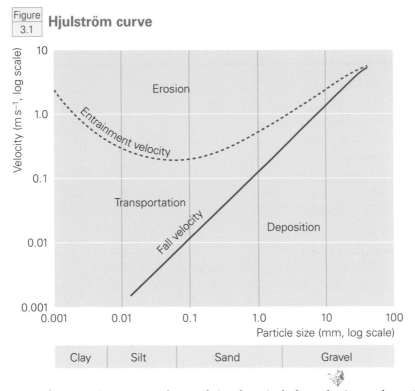

Figure 3.1 Hjulström curve

In reality, erosion cannot be explained entirely by velocity and particle size. It is influenced by a number of other factors, including:

- the effect of larger particles sheltering smaller ones
- the extent to which the stream bed is loosely or tightly packed with sediments (e.g. armoured)
- the extent to which particles protrude into the flow

Processes of erosion

Fluvial erosion consists of the processes of **corrasion** (abrasive and hydraulic action), **cavitation** and **corrosion**.

Abrasive action is the grinding effect on the bed and banks of water armed with rock particles. Coarse bedload and sand-sized particles are the main tools of abrasion. As these particles are transported, they scour the channel and undermine the banks and valley slopes. In areas where solid rock crops out, pebbles and gravels drill **potholes** in the stream bed.

Activity 1

(a) Using the data in Table 3.1, plot in Excel the critical erosion velocity curve for particles of different size. Both axes should be converted to a logarithmic scale. Sediment size will appear on the x axis; velocity on the y axis.

(b) Describe and explain the relationship between critical erosion velocity and particle size.

Table 3.1

Particle size (mm)	Velocity (cm s^{-1})
256	365.8
128	333.0
64	274.3
32	222.5
16	152.4
8	97.5
4	61.0
2	39.6
1	24.4
0.5	21.3
0.25	18.3
0.125	21.3
0.0625	24.4
0.031	33.5
0.016	48.8
0.008	67.1
0.004	106.7
0.002	167.6
0.001	243.8
0.0005	335.3
0.00024	426.7

Hydraulic action describes the erosive effect of flowing water without the assistance of rock particles. Forces of shear stress, hydraulic lift and drag are most effective where river banks are made from non-cohesive, gravelly material. Hydraulic action is often evident on meander bends, where high-velocity flow occurs close to the outer (cut) bank. Here the fast flow entrains bank material and scours the base of the bank, ultimately leading to collapse (Figure 3.2).

Cavitation is also a type of hydraulic action. It occurs when tiny bubbles of air implode in fissures and cracks in channel banks. The resulting shock waves weaken bank materials and eventually lead to collapse.

Corrosion is the chemical action of stream water, which dissolves carbonate rocks such as limestone.

Figure 3.2 **Bank collapse due to undercutting on the River Wharfe, North Yorkshire**

Table 3.2 **Factors influencing bank erosion**

Factor	Relevant characteristics
Flow properties	Magnitude–frequency and variability of stream discharge; magnitude and distribution of velocity; amount of turbulence
Bank material	Size, cohesiveness and stratification of bank sediments
Climate	Amount, intensity and duration of rainfall (wetting and drying), and freeze–thaw
Subsurface conditions	Soil moisture, seepage and porewater pressures
Channel geometry	Width, depth, slope, bank height and angle, and bend curvature
Biology	Type, density and root system of vegetation; trampling by livestock; animal burrows
Human factors	Urbanisation, land drainage, reservoir development, boating and bank protection structures

Sediment transport

Streams and rivers can be thought of as enormous conveyor belts, transporting sediment from the upper to the lower parts of catchments (see Table 3.3). The upper reaches (headwaters) are the primary sources of sediment. Here rock debris is delivered to stream channels by fluvial erosion, weathering and mass movements such as rockfalls, surface wash, landslides and mudflows.

Table 3.3 **Types of sediment load and transport processes**

Type of load	Sediment characteristics	Transport processes
Bedload	Coarse particles such as boulders, cobbles and pebbles	Intermittent sliding, rolling and hopping along the river bed at high velocity; lift and eddying are also important
Suspended or wash load	Fine particles of silt and clay, and medium-sized particles of sand	Silts and clays are entrained at high velocities and transported long distances in suspension; sand-sized particles are moved at lower velocities, bouncing off the stream bed into the current (saltation)
Solution load	Dissolved minerals from weathering and dissolution of carbonate rocks, such as chalk and Carboniferous limestone, that crop out in the channel	Minerals in solution; occurs continuously and is independent of velocity

The amount of sediment transported by a river is known as the **load**. The sediment load can be sub-divided into three fractions: **bedload**, **suspended** or **wash load** and **solution load**. The size distribution of particles transported by rivers is known as the **calibre** of the load. The **competence** of a stream or river is the maximum size of sediment it can transport (Figure 3.3).

Figure 3.3 **Sediment budget**

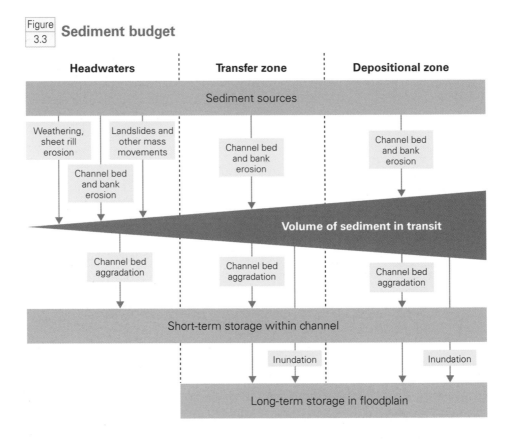

Transport processes

Streams transfer sediment by the processes of **traction**, **saltation**, **suspension** and **solution**.

Traction

Traction is the process that transports coarse bedload particles (Figure 3.4). Video evidence shows that bedload particles move intermittently by sliding, rolling and hopping along the stream bed. Movement is intermittent because of:
- unevenness in the shape of the stream channel
- turbulent flow, with localised high velocities, eddies and vortices

Figure
3.4
Coarse bedload or lag deposits, Harwood Beck, Upper Teesdale

Traction is closely linked to turbulent flow conditions and shear stress. The frictional resistance of the channel bed creates a zone of slow water along the bottom. Above the stream bed, the velocity of flow increases rapidly. This creates conditions of shear stress. The momentum of this faster moving water is transmitted to the slower moving boundary water, which is 'rolled up' in a spiral motion. This process is responsible for the movement of bedload particles in a rolling and sliding motion downstream (Figure 3.5).

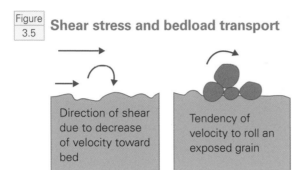

Figure 3.5 **Shear stress and bedload transport**

Direction of shear due to decrease of velocity toward bed

Tendency of velocity to roll an exposed grain

Traction inevitably leads to the collision of particles and abrasion of the channel. As a result, boulders, cobbles and gravels are chipped and split and become smoother and more rounded. This process is called **attrition**.

Bedload transport can only happen when stream energy levels are high. Indeed, the largest particles may only move at times of bankfull. Studies of the movement of bedload have shown that, on average, particles travel only short distances — less than four channel widths during each high-flow event.

Saltation

Saltation is a skipping motion of sand-sized particles (Figure 3.6). Individual sand grains bounce along the river bed in an arc-shaped trajectory and are carried downstream by the current. The process is cumulative. Once initiated, the impact of saltating grains sets in motion more and more sand particles on the river bed.

Figure 3.6 **Saltation and the transport of sand-sized particles**

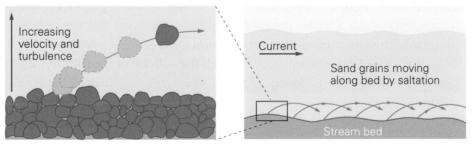

Increasing velocity and turbulence

Current

Sand grains moving along bed by saltation

Stream bed

Suspension

Only the smallest particles — silt and clay — are transported in suspension. Although it takes high velocities to entrain clay and silt particles, once in suspension they are transported long distances (Figures 3.1. and 3.7).

Figure 3.7 **Suspended sediment load in a stream channel at Repton, Derbyshire**

Solution

Streams draining areas of chalk and limestone transport weathered minerals in solution. In the humid tropics, where the chemical weathering of rocks is highly efficient, the solution load is important. Unlike the bedload and suspension (or wash) load, transport of minerals in solution takes place at frequent intervals, when flow is well below bankfull. Although the volume of the solution load is far less than traction and suspended load, it is more significant than is often assumed.

Downstream fining of sediments

Stream sediments tend to get finer with increasing distance downstream. Headwater streams are dominated by boulders and cobbles; in lowland streams, sand, silt and clay particles are more frequent. We refer to this trend as the downstream fining of sediments (Figure 3.8).

Activity 2

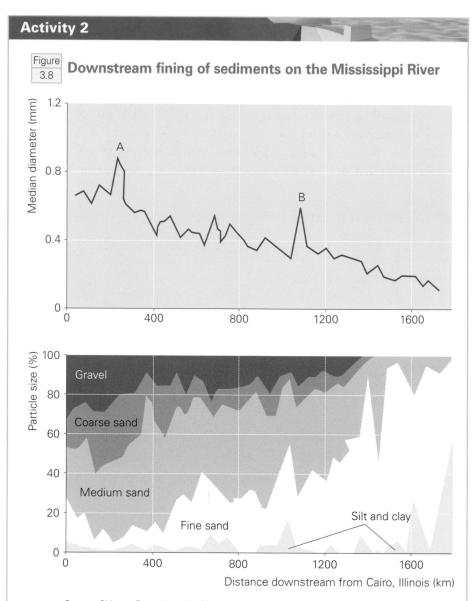

Figure 3.8 Downstream fining of sediments on the Mississippi River

Source: Skinner, B. et al. (1999) *Blue Planet: Introduction to Earth System Science*, John Wiley.

Study Figure 3.8, which shows downstream changes in sediment size on the Mississippi River between Cairo, Illinois, and the mouth of the river.

(a) Describe and suggest two possible explanations for the downstream changes in sediment size on the Mississippi River.

(b) Suggest a possible reason for the increases in sediment size at locations A and B.

Downstream fining is partly due to the progressive abrasion and attrition of particles as they move downstream. However, this is not the only explanation. We must also take into account the selective erosion and transport of sediment by size. The Hjulström curve (Figure 3.1) describes the selective erosion and transport of stream sediments. Medium-sized particles are easily removed from headwater reaches. Finer sediments, once entrained, settle out of suspension only at low velocities and are therefore transported long distances. This selective erosion and transport partly explains the dominance of coarse-channel sediments (lag deposits) upstream and finer particles downstream (Figure 3.9).

| Figure 3.9 | **Grain size and distance transported** |

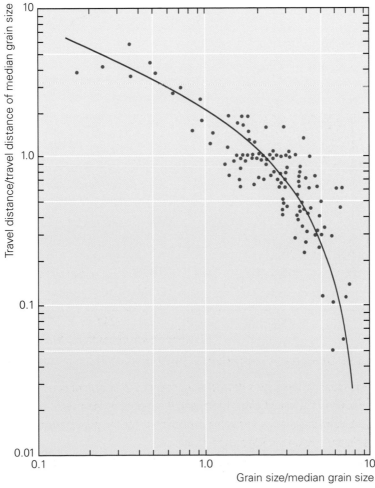

Source: Knighton, D. (1998) *Fluvial Form And Process*, Arnold.

However, the relationship between sediment size, roundness and distance downstream is complex. Tributary streams constantly input fresh sediments. Where these streams drain upland catchments, sediment inputs are often coarser and more angular than those in the main river. Contrasts in the geology of tributary catchments, and sediment inputs from bank erosion, further complicate the relationship.

Downstream fining of sediments on the River Wharfe

The bedload of the River Wharfe in North Yorkshire is mainly made up of Carboniferous limestone and sandstone. These two rock types represent the geology of the catchment upstream from Wetherby. Most of the limestone particles originate from the catchment headwaters above Burnsall, in upper Wharfedale and Littondale.

Figure 3.10 **Wharfedale**

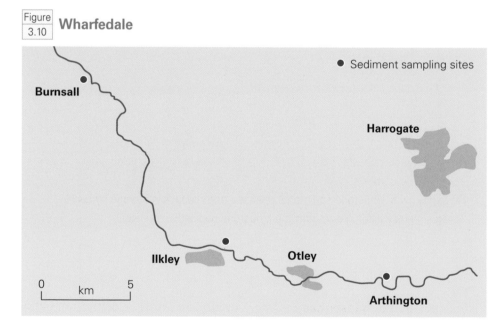

Large inputs of sandstone occur between Burnsall and Ilkley (Figure 3.10). This explains the increase in the mean size of sandstone particles between Burnsall and Ilkley shown in Table 3.4. Large inputs of sandstone downstream of Burnsall are also confirmed by the ratio of sandstone to limestone. At Burnsall, the ratio is 1:5; at Ilkley and Arthington it is almost 1:1 (Table 3.4).

Limestone particles decrease in size by 3% between Burnsall and Ilkley, and by 37% between Ilkley and Arthington. However, we cannot assume that the particles at Arthington have been transported recently from the upper catchment. This is because the River Wharfe's floodplain acts as a long-term sediment store. In places the river is actively eroding its banks, releasing limestone sediments laid down thousands of years ago when the energy environment of the river was very different from that today.

Table 3.4 Downstream changes in sediment size on the River Wharfe point bars (median axis)

Site	Statistic	Sandstone	Limestone
Burnsall	Mean (cm)	4.52	4.37
	Standard deviation (cm)	1.49	1.35
	Coefficient of variation (%)	33.00	29.40
Ilkley	Mean (cm)	5.90	4.24
	Standard deviation (cm)	2.69	1.86
	Coefficient of variation (%)	45.50	43.77
Arthington	Mean (cm)	3.10	2.67
	Standard deviation (cm)	1.51	1.38
	Coefficient of variation (%)	48.6	51.7

Activity 3

(a) Suggest reasons why, at each site in Table 3.4, sandstone particles are larger than limestone particles.

(b) Explain why limestone sediments released from floodplain storage by bank erosion may differ in size from the limestone bedload in the river.

Deposition

The deposition of transported sediment occurs once flow velocities fall below the settling velocities for particles of given size.

Settling velocities are closely related to particle size (Figure 3.1). As energy levels decline, the coarsest particles are deposited first, followed by progressively finer ones. The outcome is a sorting of sediments by size, both horizontally and vertically. For example, point bar deposits often show horizontal size sorting, with fining occurring with distance from the low-flow channel. In contrast, floodplain deposits often develop vertical sorting. The lowest layers are lag particles, originally formed as sediment bars within the channel.

They are usually topped-off with **overbank sediments** of fine silt and clay deposited in times of flood (Figure 3.11). The process of sediment deposition within a channel is known as **aggradation** and, as we shall see in Chapter 5, is responsible for a variety of depositional landforms.

Figure 3.11 **Suspended sediment deposits overlying gravels exposed on a cut bank, River Wharfe, Arthington**

4 Channel patterns

Channel pattern or **planform** describes the shape of a stream on a map or aerial photograph. Just like width, depth, velocity and gradient, channel pattern is a dependent variable that adjusts to changes in discharge and sediment load.

| Figure 4.1 | **Types of channel pattern** |

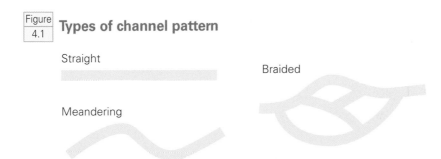

Straight

Meandering

Braided

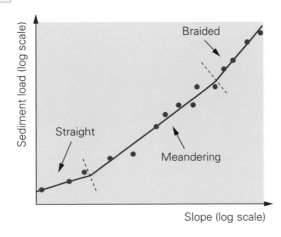

At the simplest level, channel patterns fall into two types — single-thread and multi-thread. Within this classification there are straight, meandering and braided patterns (Figures 4.1, 4.9, 4.11, 7.12). In reality, there are several intermediate types, often determined by the characteristics of the sediment load (Figures 4.2 and 4.3). As with sediment load, channel patterns are also influenced by discharge, slope and bank materials.

| Figure 4.2 | **Channel pattern, gradient and sediment load** |

Figure
4.3

Channel patterns and sediment load

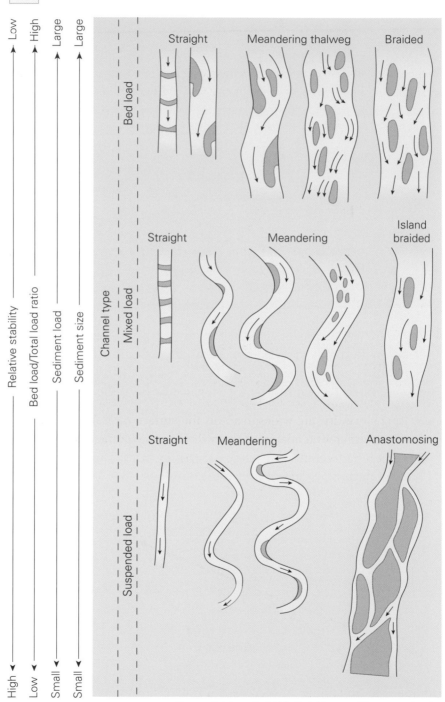

Source: Knighton, D. (1998) *Fluvial Form and Processes*, Arnold.

Straight channels

Natural river channels are rarely straight for a length greater than ten times their width. Straight channels are associated with low-energy streams that have limited power for transportation and erosion. Fault-guided streams or streams flowing in bedrock channels are also often straight. Straight channels are found immediately downstream of waterfalls too. However, when streams are artificially confined to straight channels (see Chapter 7), and where bank materials are erodible, they adjust quickly and establish more sinuous planforms.

Meandering channels

Sinuous stream channels are also known as meandering channels. Strictly speaking, a meandering pattern is one in which the ratio of channel length to valley length (sinuosity) exceeds 1.5 (Figure 4.4). Meandering channels are usually:

- single thread
- developed in coherent bank material such as silt and clay
- formed by streams with relatively high energy

Most streams and rivers show some degree of meander. Indeed, meandering channels eroded by water are not confined to Earth — ancient meandering channels, long since dry, are widespread on the surface of Mars. Meandering appears to be an inherent tendency of water flowing in channels and is related to the interaction between flowing water and the boundary surfaces.

| Figure 4.4 | **Measuring sinuosity** |

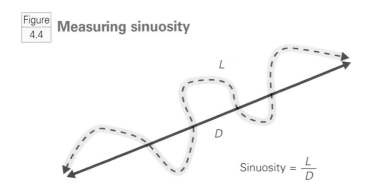

$$\text{Sinuosity} = \frac{L}{D}$$

One approach to explaining meanders is to think of streams as energy systems. Streams with coherent (but erodible) bank materials, shallow gradients and excess energy will expend this energy on transport and lateral erosion. Lateral erosion increases sinuosity, thereby lengthening the channel. This reduces the average gradient and dissipates excess energy. In this way, equilibrium is restored to the energy budget of a stream. However, it is clear that in the first instance a stream with a meandering channel must have sufficient power to erode its banks.

Formation of meanders

Meanders are so widespread and regular in nature that they are unlikely to result from simple random disturbances. However, there is, at yet, no single or completely satisfactory explanation of meander formation. Although in general terms meandering channels are related to the interaction between flowing water, channel boundaries and mobile bed forms such as bars, meanders also develop where there are no confining channel boundaries and no sediment. Examples include the meanders of the Gulf Stream and North Atlantic current, and meltwater streams on glacier surfaces where sediment is largely absent.

Meanders and channel sediments

Even in straight channels the path of maximum velocity (**thalweg**) has a sinuous course. In these circumstances, **alternating bars** of sediment form by deposition in the lower velocity areas on either side of the thalweg and extend downstream (Figure 4.5). These bars deflect the flow against the opposite bank, initiating erosion, channel curvature and the formation of meanders.

Bars of coarse sediment also form at the points of inflection in sinuous channels, where energy levels are low. These bars increase channel roughness, which causes further accretion. Depleted of its load, the stream now has surplus energy and scours the bed to form a pool downstream in the incipient meander. The eroded material is then deposited on the next bar. This pattern of alternating erosion and deposition produces a sequence of **pools** and **riffles**, spaced uniformly at an average distance of 5 to 7 times channel width.

In fully developed meanders, pools are located at the apex of curvature, close to the **cut bank** where flow velocity is highest (Figure 4.5). These pools are areas of scour, laminar flow, deep water and fine sediment. Riffles occur on straight sections of channel where the thalweg swings from one bank to the other. They are areas of shallow water, coarse sediments, steep gradient and turbulent flow.

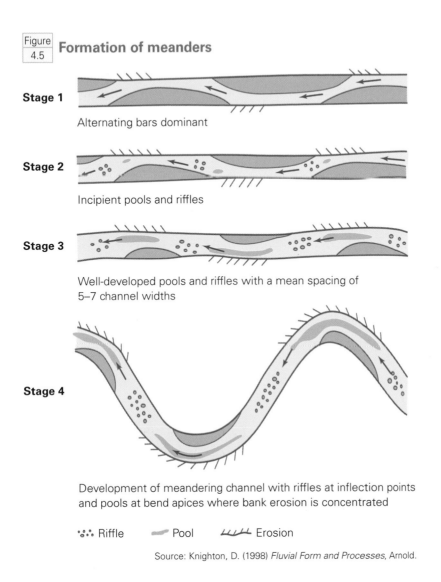

Figure 4.5 **Formation of meanders**

Stage 1

Alternating bars dominant

Stage 2

Incipient pools and riffles

Stage 3

Well-developed pools and riffles with a mean spacing of
5–7 channel widths

Stage 4

Development of meandering channel with riffles at inflection points
and pools at bend apices where bank erosion is concentrated

•:•. Riffle   Pool   ∠∠∠∟ Erosion

Source: Knighton, D. (1998) *Fluvial Form and Processes*, Arnold.

Flow patterns in meanders

Distinctive flow patterns develop in meandering channels. In addition to the
main downstream flow, there is a secondary spiral or **helical flow** (Figure 4.6).
As the flow enters a meandering channel, centrifugal forces cause the water to
pile up against the outer cut bank. The elevated water surface produces a
hydraulic gradient that acts towards the inner bank and creates a current at depth
from the outer to the inner bank. This current balances the surface flow towards
the cut bank and completes the secondary circulation. As this current weakens
towards the inner bank, it deposits sediment to form a point bar.

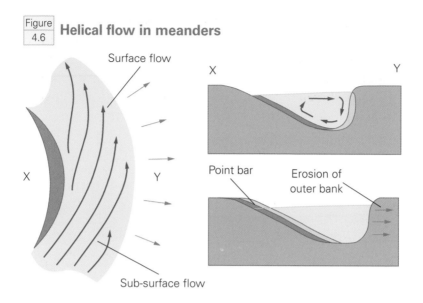

Figure 4.6 **Helical flow in meanders**

Meander geometry

Meanders often show remarkable symmetry (Figure 4.7) that can be compared to sine-generated curves.

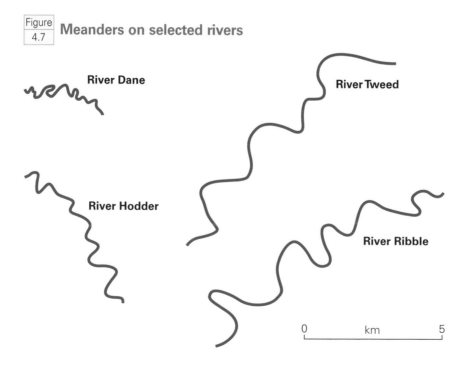

Figure 4.7 **Meanders on selected rivers**

A number of measures have been devised to describe meander size and geometry (Figure 4.8). Strong relationships have been found between some of these measures. For example, meander wavelength has a strong positive correlation with bankfull discharge and channel width (wavelengths are usually around 7 to 10 times channel width). The strength of these relationships provides proof that meandering channels are equilibrium forms.

Figure 4.8 **Meander geometry**

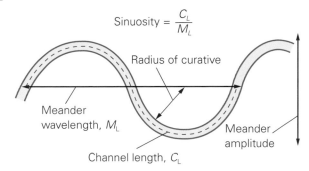

$$\text{Sinuosity} = \frac{C_L}{M_L}$$

Radius of curative

Meander wavelength, M_L

Channel length, C_L

Meander amplitude

Figure 4.9 **Point bars and alternate bars on the River Skirfare**

Activity 1

(a) Using Figure 4.7, measure the average sinuosity, average wavelength and average amplitude of the Rivers Tweed, Ribble, Hodder and Dane.

(b) Investigate the relationship between discharge and meander wavelength and amplitude for the Rivers Tweed, Ribble, Hodder and Dane by plotting the data on discharge given in Table 4.1 and your measurements of average wavelength and average amplitude as two scattercharts showing the relationships.

| Table 4.1 | Average discharge on the Tweed, Ribble, Hodder and Dane |

River	Mean discharge (cumecs)
Tweed	78.87
Ribble	33.23
Hodder	8.55
Dane	2.39

(c) Comment on the results.

Activity 2

Study Figures 4.5 and 4.9.

(a) Make a sketch of the stream channel shown in Figure 4.9 and label the following features: point bar, alternate bars, pools, riffles and areas of erosion.

(b) Compare the planform of the stream in Figure 4.9 downstream of the meander with Figure 4.5. State and justify its current stage of development.

(c) On your sketch, show how the stream channel is likely to change its position and planform in future.

(d) Explain the changes you described in (c).

Braided channels

Braided channels are multi-thread, with individual channels separated by bars or islands. In braided channels, these bars are unvegetated and submerged at high flow. Typical braided channels are relatively shallow and wide and have high width-to-depth ratios. This results in turbulent flow that makes braided channels effective for transporting large volumes of coarse sediment. Unlike meandering channels, braided channels are unstable and shift position constantly.

Braiding is caused by:

- rapid bank erosion
- overloading the channel with sediment

Conditions favouring braiding

Braided streams have considerable power. This power comes from steep slopes (and therefore high velocity flow) and seasonally high discharge. With large amounts of surplus energy, bank erosion is rapid.

Braiding is found in many different environments, from glaciated regions to deserts. In southern Iceland, the streams draining south from Vatnajökull and other ice fields have well-developed braided channels. This braiding is associated with:

- rapid and frequent variations in discharge, especially meltwater in spring and early summer
- huge volumes of coarse glacial debris, which often exceed the transport capacity (competence) of the rivers
- highly erodible gravel banks, which also contribute to high sediment loads
- a lack of vegetation cover, increasing the erodibility of channel banks
- steep channel gradients

Development of braided channels

Because braided streams usually have easily erodible banks made of sands and gravels, they have considerable freedom to erode laterally. This allows their channels to adjust to variable discharge and sediment inputs. It is the exact opposite of mountain streams, which are constrained by narrow, rocky channels and to a lesser extent of meandering streams, which are confined by coherent banks of silt and clay.

Braiding often begins with the deposition of coarse **lag material** in the channel. This forms a central or transverse bar in the stream channel and acts as a focus for further deposition. As the bar grows both vertically and downstream, the main flow is diverted to the smaller channels on either side and causes accelerated bank erosion. Eventually this leads to further channel deposition and the process begins anew (Figure 4.10). Braiding may also develop when point bars and alternate bars are dissected by **chute cut-offs** to form **medial bars**.

Figure
4.10

The process of braiding

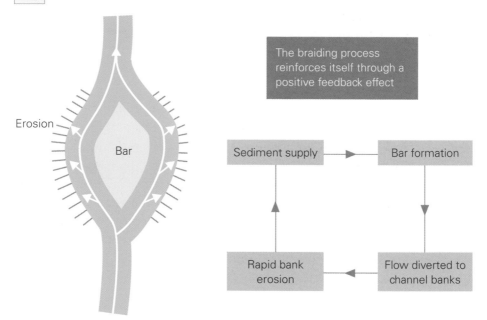

Erosion

Bar

The braiding process reinforces itself through a positive feedback effect

Sediment supply → Bar formation

Rapid bank erosion ← Flow diverted to channel banks

Figure
4.11

Braiding at Boverdal, Norway

Braided channels tend to shift position rapidly. However, despite their instability, braiding channels probably represent an adjustment made by powerful rivers which have erodible banks and sediment loads too large to be carried in a single channel. In other words, braided channels — like meandering channels — are essentially an equilibrium form.

Anastomosing channels

Anastomosing streams, like braided streams, are multi-channelled. However, they differ in a number of ways. First, anastomosing streams are less powerful and have shallower gradients. Second, they have cohesive banks that are not easily eroded. Third, they have relatively narrow and deep channels with a low width-to-depth ratio. Aggradation is the dominant process in anastomosing channels. This may be due to a rising base level causing a rapid build-up of sediment, or to a decrease in energy (e.g. where a stream enters a lake or flows from a steep, narrow valley onto an alluvial fan) — situations in which single-thread channels easily break up into multi-thread channels.

5 Fluvial landforms

The work of streams and rivers as agents of sediment transport was used by Schumm (1977) to divide drainage basins into:

- the sediment supply zone
- the sediment transfer zone
- the sediment storage zone

Within each zone, a combination of processes gives rise to distinctive channel and valley **landforms**.

In large drainage basins, the sediment supply zone is usually a mountainous or upland region. The sediment storage zone lies close to sea level and often includes a lowland coastal plain and a delta. Between the upland and lowland zones there is an intermediate zone where transport processes dominate. Here, inputs and outputs of sediment are roughly equal. Even so, sediment may be stored in the transfer zone in floodplains and terraces for long periods of time.

Schumm's model is a useful description of the spatial distribution of fluvial processes and landforms. However, it is an oversimplification. While a single fluvial process may dominate a particular zone, erosion, transport and deposition are found in all three zones.

The sediment supply zone

Sources of sediment

In any drainage basin, the headwater zone is the main source of sediment. A number of factors combine to produce high rates of sediment supply in this zone, including:

- sparse vegetation cover (e.g. lack of woodland), which contributes to mass movement and physical weathering on slopes
- steep valley slopes, which trigger mass movements such as landslides and earthflows
- high drainage density, which transfers sediment rapidly to stream channels
- climate extremes, with frequent freeze–thaw cycles and high-intensity precipitation, which intensify rock weathering
- steep channel gradients and high elevations, which create surplus stream energy for channel incision

Headwater streams expend surplus energy on sediment transport and erosion. Meanwhile, the nature of sediment input — coarse gravels and cobbles — provides them with the tools for abrasion, which in turn generates further sediment input.

Most sediments in the headwater zone are stored within stream channels as **alluvial bars**. These are temporary stores, which often remain *in situ* only until the next bankfull event.

Landforms in the sediment supply zone

V-shaped valleys

Valleys in headwater regions are typically steep-sided, narrow and V-shaped in cross-section (Figure 5.1).

| Figure 5.1 | **Formation of V-shaped valleys** |

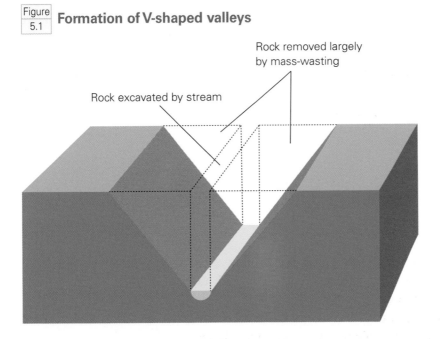

Rock removed largely by mass-wasting

Rock excavated by stream

These valleys owe their form to rapid fluvial incision and the back-wasting of valley slopes. Confined within resistant rock walls, lateral erosion is limited and erosion is mainly vertical. As the channel becomes incised, the valley sides, particularly on more sinuous reaches, are undercut and become unstable. Slope failure results in mass movements such as rotational slides and rockfalls. The outcome is a lowering of valley-slope angles, valley widening and the formation of classic V-shaped cross-sectional profiles (Figures 5.2 and 5.3).

Figure 5.2 **(a) Valley slope failure caused by undercutting and (b) sediment input from weathering and erosion of valley slopes, Marshaw Wyre, Lancashire**

Most upland streams show some degree of sinuosity. This characteristic, combined with rapid down-cutting, results in the formation of interlocking spurs (Figure 5.3).

Figure 5.3 **V-shaped valley with interlocking spurs: Upper Nidderdale, North Yorkshire**

The long profiles of headwater streams are typically irregular, reflecting the influence of geology and rock structure. Solid rock frequently crops out in stream channels, and more resistant bands form rock steps and waterfalls that retreat gradually upstream.

Step–pool sequences

Many high-gradient, boulder-bed streams have channels with alternating steep (step) and gentle (pool) segments. Cobbles and boulders form the steps, which alternate with finer sediments in pools to produce a repetitive, stair-like long profile in the stream channel (Figures 5.4 and 5.5).

| Figure 5.4 | **Step–pool sequences, River Traligill, Sutherland** |

| Figure 5.5 | **Longitudinal profile of a step–pool sequence** |

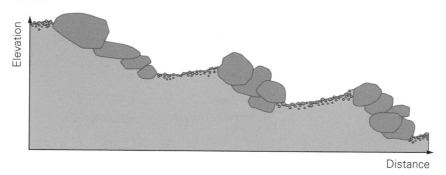

Step–pool sequences are an adjustment in a vertical dimension by mountain streams, which are normally confined by narrow rocky valleys (Table 5.1). Unlike low-gradient alluvial streams, mountain streams cannot adjust to changes in discharge, sediment load and the energy system by moving laterally (e.g. meandering). Adjustment is therefore limited to the vertical dimension. It is thought that steps, by increasing flow resistance, allow surplus energy (heat energy) to be dissipated as turbulent flow.

Table 5.1	**Stages in the development of a step–pool sequence**

Stage	Description
Stage 1	▪ River transports cobbles and boulders at high flow ▪ As energy levels fall, these aggregate, forming a step in the bed
Stage 2	▪ Turbulent flow occurs over the step, dissipating energy because of increased roughness ▪ At high flow, scouring (abrasion) occurs in the area immediately downstream of the step, forming a pool
Stage 3	▪ Sediment scoured from the pool aggregates on the downstream step
Stage 4	▪ The stair-like profile results in turbulent flow ▪ Loss of energy on the steps leads to equilibrium conditions, with the stream having just enough energy to transport its load

Periodicities have been identified in the distribution of steps and pools in mountain streams. Recent work suggests that they are spaced at regular intervals along the channel, at 0.4–2.4 times channel width. The distance between successive steps or successive pools is known as the wavelength, and is similar to pool–riffle sequences found in low-gradient streams. Such periodicity is strong evidence of an adjustment between flow and channel morphology.

Activity 1

Table 5.2	**Ratio of pool wavelength to median bankfull width, Backstone Beck, West Yorkshire**

	Wavelength-to-bankfull width ratio					
	0–0.99	1–1.99	2–2.99	3–3.99	4–4.99	5–5.99
Observed	1	18	14	4	2	1
Expected	6.66	6.66	6.66	6.66	6.66	6.66

Table 5.2 shows the ratio of pool wavelength to median bankfull channel width for a 150 m reach of Backstone Beck, West Yorkshire. Backstone Beck is an upland stream, with steep channel gradients and coarse, poorly sorted bedload.

(a) Use the chi-squared test to test the hypothesis that the distribution of observed ratios is no different from an expected random distribution.

(b) Explain the result of the chi-squared test.

Activity 2

Investigate a 100–200 m step–pool sequence through fieldwork.

(a) Make a risk assessment of the site (hazards, safety precautions, accessibility etc.).

(b) Make notes on the character of the stream and its channel, for example size and sorting of bedload, channel gradient, and presence or absence of solid rock in the channel.

(c) Use a tape to measure wavelengths — the distance between successive steps in the channel — and calculate the median bankfull channel width for each wavelength. (For each wavelength, measure five bankfull channel widths and calculate the median width.)

(d) Present the data as:
  - a frequency table (as Table 5.2)
  - a histogram

Test for significance using the chi-squared test.

Alluvial bars

With floodplains absent from the headwater zone, long-term sediment storage is minimal. Instead, as energy levels subside following bankfull discharge, sediments are deposited within channels as alluvial bars. This within-channel deposition is known as **aggradation**. Alluvial bars include **mid-channel bars**, **alternate bars** and **point bars**. These have a strong influence on channel type and are responsible for braided and step–pool sections.

Headwater streams with large, coarse sediment loads are most likely to show alluvial bar development. This can be explained by:

- the greater frictional resistance to sediment transport provided by coarse sediments
- shallow and inefficient stream channels
- the relatively small reductions in flow velocity needed to cause deposition of the bed load (Figure 3.1).

Mid-channel bars

Mid-channel bars have long axes parallel to stream flow, hence their alternative name of longitudinal bars. They develop around a nucleus of coarse sediment. As discharge and flow competence decline, coarse sediments accumulate on the upstream side of this nucleus. Here they become tightly packed and difficult to dislodge (Figures 5.6, 5.7 and 5.8).

Figure 5.6

Mid-channel bar formation

Stage 1

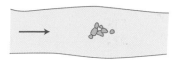

Coarse sediment is deposited in mid-channel as flow competence declines

Stage 2

Obstacle to flow promotes further aggradation, and deflects flow against banks

Stage 3

Bank erosion releases sediment; growth of mid-channel bar and shallowing of channel

Stage 4

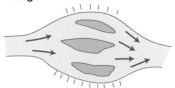

Bar stabilises with deposition of fine sediment and growth of vegetation

Figure 5.7

Mid-channel bar, Kingsdale Beck, North Yorkshire

Erosion and undercutting on outside of meander releases coarse sediment into the channel

Aggradation steepens channel gradient, increasing stream energy and competence to transport sediment

Aggradation of eroded sediment in mid-channel as flow competence declines

Figure 5.8 **(a) Sediment packing; (b) particle size distribution, Cowside Beck, North Yorkshire**

(a)

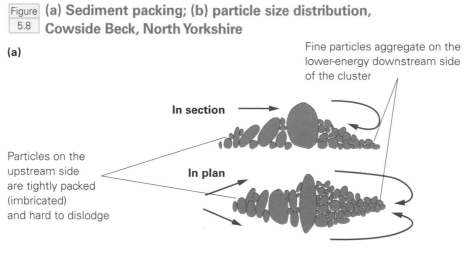

Particles on the upstream side are tightly packed (imbricated) and hard to dislodge

Fine particles aggregate on the lower-energy downstream side of the cluster

In section

In plan

Pebble clusters form the nucleus of mid-channel and alternate bars

(b)

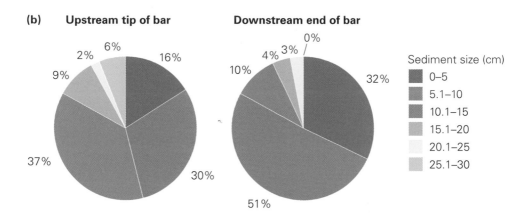

Upstream tip of bar

6%
2%
9%
16%
37%
30%

Downstream end of bar

0%
4% 3%
10%
32%
51%

Sediment size (cm)
- 0–5
- 5.1–10
- 10.1–15
- 15.1–20
- 20.1–25
- 25.1–30

Finer material is then deposited on the downstream side (Figure 5.8). As a mid-channel bar grows and the channel bifurcates, stream flow is deflected towards the banks, which erode and release further sediment into the channel. The subsequent reduction in channel efficiency has a positive feedback effect, causing further aggradation and bar development. Ultimately this process leads to the formation of braided channels. Some mid-channel bars, topped by fine sediments and colonised by plants, become permanent features, only submerged during periods of exceptionally high flow.

Alternate bars

Alternate bars form on opposite sides of straight channels (Figure 5.9).

Figure 5.9 **Formation of alternate bars**

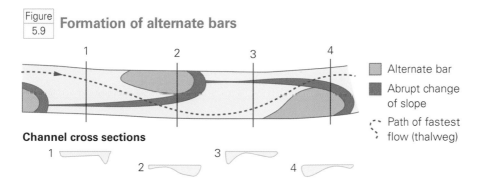

Channel cross sections

The thalweg meanders between alternate bars, which develop in the lower energy areas along the sides of straight channels in high-flow conditions. At low flow, the bars are exposed. The channel bed often has an abrupt change of slope on the downstream side of the bar, which gradually migrates upstream (Figures 5.10 and 5.11). We have seen that alternate bars are an early stage in the formation of meandering channels (Chapter 4).

Figure 5.10 **Abrupt change of gradient on an alternate bar, River Skirfare, North Yorkshire**

Figure
5.11 **Alternate bars on the River Skirfare, North Yorkshire**

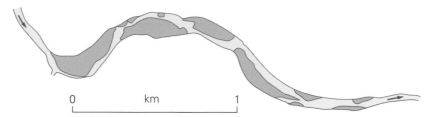

0   km   1

Alluvial bars and stream equilibrium

The formation of alluvial bars is another example of how streams restore equilibrium by adjusting their channel gradient. Aggradation steepens the channel gradient and increases flow velocities and energy levels, re-establishing the balance between the input and output of sediment (Figure 5.12).

Figure
5.12 **The relationship between channel gradient and channel width**

Activity 3

Select a 200 m section of channel in the headwater reaches of a small stream. The section should be developed on alluvium and include some variation in bankfull channel width and channel gradient.

(a) Measure 20 channel cross-sections at equal intervals along the channel. At each cross-section measure:
- bankfull width
- average bankfull depth
- average gradient over a distance of 5 m upstream and 5 m downstream from each cross-section
- wetted perimeter of each cross-section

(b) Plot the long-profile gradient of the channel over the 200 m section.

(c) Use the Spearman rank correlation to test the following relationships:
- channel width and gradient
- width:depth ratio and gradient
- hydraulic radius and gradient

(d) Summarise your findings and evaluate your methodology and outcomes.

Activity 4

(a) Explain the relationships between channel gradient, channel width, width:depth ratio and hydraulic radius given in Figure 5.12.

(b) Using the data in Table 5.3, plot scattergraphs to show the relationship between:
- channel gradient and channel width
- channel gradient and width:depth ratio

Table 5.3 Channel relationships: nine reaches on the Marshaw Wyre

Width (m)	Width:depth ratio	Gradient (°)
5.65	10.29	2.51
6.69	14.50	1.37
7.30	16.04	3.27
4.90	7.34	1.40
13.71	43.68	3.58
4.29	7.07	1.80
5.39	10.59	1.13
9.42	25.91	2.30
7.02	20.6	2.33

(c) For the two data sets, calculate the Spearman rank correlation coefficients and their statistical significance. Comment on your results.

The sediment transfer zone

Landforms in the sediment transfer zone

Streams emerging from their headwaters enter the sediment transfer zone. There are significant differences between the headwater zone and the sediment transfer zone. In the transfer zone:
- sediment inputs from erosion and mass movements are lower
- sediment storage times are longer, with much of this storage in floodplains
- channel gradients are gentler and are developed on alluvium, allowing greater lateral movement of stream channels and meander development
- average velocities are often higher because of the greater efficiency of channel shape and finer sediments, which offer less resistance to flow

Point bars

Point bars are semicircular sand or gravel deposits located on the inner banks of meanders (Figure 5.13). They are depositional features whose development is linked to the spiral flow of water in meanders, known as helical flow (Figure 4.6).

Figure 5.13 **Point bar deposits, Marshaw Wyre, Forest of Bowland**

We saw in Chapter 4 that helical flow raises the water level on the outer bank and produces a transverse current towards the inner bank, close to the bed and at an angle to the main current. This transverse current decreases in strength towards the inner bank and up the point-bar surface. Deposition occurs on the point bar when the energy to transport individual particles no longer exceeds particle weight.

The outcome of this process is the sorting of sediments on point bars (Figure 5.14). This sorting is evidence of the transverse current and the loss in energy towards the inner bank. Larger particles are deposited on the lower surface of the point bar. The gradual dissipation of energy leads to progressive fining of sediment towards the upper surface. The effect of helical flow is also shown by the orientation of disc-like particles, whose long axes are parallel to the flow.

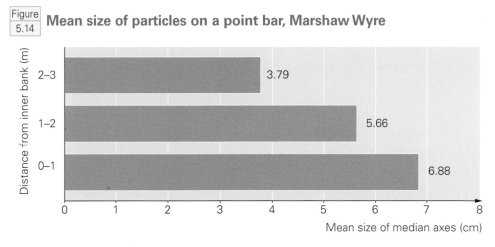

Figure 5.14 Mean size of particles on a point bar, Marshaw Wyre

As river channels shift laterally, erosion on the outside of a meander is compensated by deposition on the inside. Thus point bars gradually extend streamwards and increase in height. Eventually, point-bar deposits are incorporated into the floodplain of the river.

Figure 5.15 shows the sequence of deposition on the point bar of a meandering channel as the channel shifts position over time. T_1–T_3 are time intervals and show how the point bar deposits, sorted by size, build up the floodplain as the channel migrates across the valley floor.

Figure 5.15 Formation of point bars

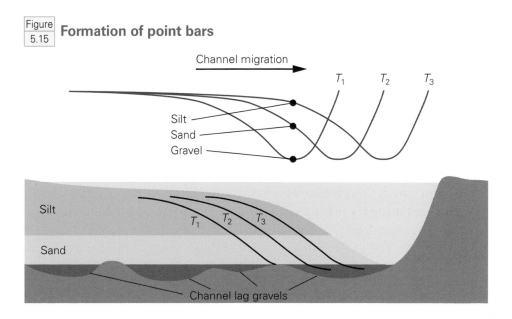

Activity 5

Analyse point-bar deposits for evidence of sorting and helical flow. Use the following methodology in the field:

- Identify three or more semicircular zones on the point-bar surface, parallel to the low-flow channel. Each zone should have the same width.
- Locate the approximate centre of the radius of curvature on the point bar. Using a tape, establish line transects from this point to the edge of the low-flow channel.
- Collect a sample of particles at regular intervals along each transect. For each particle, measure its median axis and identify its zonal location. (A possible extension of the study would include the orientation of the long axes of the particles.)
- Present the data using appropriate charts and use chi-squared to test for statistical significance.
- Evaluate your methodology and outcomes.

Pool and riffle sequences

In a vertical dimension, the most characteristic channel landforms in the transfer zone are pools and riffles (Figure 5.16). Like step–pool sequences, pools and riffles are related to stream energy expenditure. However, unlike step–pool sequences, pools and riffles are developed in alluvial (rather than in boulder-bed) streams and on gentler slopes.

| Figure 5.16 | **Pool–riffle sequences** |

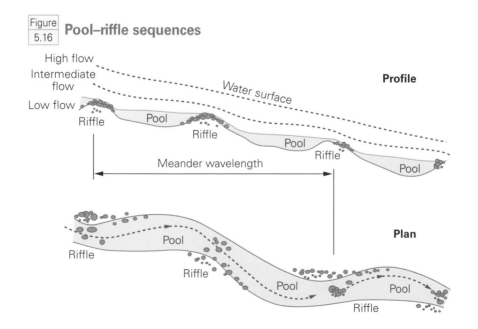

Pools and riffles develop in both straight and meandering channels. Pools are areas of scour, fast flow and fine sediment. Riffles are lobes of gravel, exposed at low flow, and are aggradational features. They have steeper slopes, coarser sediments and shallower flow depth than pools. The spacing of pools and riffles is regular — on average, 5 to 7 times channel width. This spacing is similar to the ratio between meander wavelength and channel width and suggests that pool–riffle sequences play a part in meander formation (Figure 4.5).

There are several explanations of pool–riffle sequences. The most plausible theory is that pools and riffles are a way in which alluvial streams dissipate excess energy. Most of the surplus energy is expended on scouring the pools at high flow. Some energy is also lost through turbulent flow across the riffles.

Alluvial fans

Alluvial fans are cones of sediment deposited by streams (Figure 5.17). They often form where streams emerge from upland valleys confined by steep rock walls into areas of low relief. These streams, with high velocity and high energy, often transport large sediment loads.

Figure 5.17 **Alluvial fan development and form**

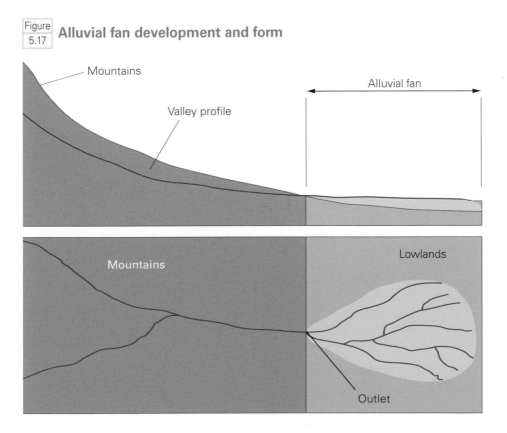

As streams emerge from upland to lowland environments they experience a sudden loss of energy. No longer confined by rock walls, the stream channels become shallower and wider and this causes an increase in resistance to flow. The result is channel aggradation, the formation of multiple-thread channels, and alluvial fans.

- Alluvial fans form where streams and rivers leave mountainous courses for the lowlands.
- The abrupt change in gradient and expansion of the valley at the mountain–lowland boundary causes rapid deposition.
- Debris flow may deposit sediment at the upstream end of the fan.
- Sediment becomes finer towards the downstream end of the fan, where sheetflow may occur.

Alluvial fans commonly occur in the British Isles in upland regions where narrow tributary valleys enter larger valleys, such as glacial troughs. In the USA, alluvial fans are widespread in the Great Basin in southwest USA. Here, mountains alternate with down-faulted basins to form classic basin-range topography. Where streams enter the basins from the surrounding mountains, they build extensive debris cones (Figure 5.18). Eventually these cones merge to form **bajadas** — vast aprons of alluvium which fringe the lower slopes of the mountains.

The foreground in Figure 5.18 is an alluvial fan at the foot of a heavily eroded mountain range.

| Figure 5.18 | **Alluvial fan, Death Valley, California** |

In northern Venezuela, the Costa Mountains (2500 m) come within 15 km of the coast. Streams draining from the mountains into the sea have exceptionally steep long profiles and high energy, and have cut deep valleys. Where the streams emerge onto the narrow coastal plain, a series of alluvial fans has formed. Although exposed to hazardous debris flows, these fans have attracted several large urban developments.

Low-energy floodplains

Floodplains are landforms of both sediment transfer and sediment storage zones. Active floodplains are valley floor areas on either side of stream channels that are temporarily inundated during periods of high discharge.

Floodplain deposits

Low-energy floodplains are associated with single-thread meandering stream channels (Figure 5.19). They consist of alluvial deposits — gravels, sand, silt and clay — which often show vertical sequencing. At the base, coarse gravels dominate the sequence. They are succeeded by progressively finer deposits towards the floodplain surface.

| Figure 5.19 | **Low-energy floodplain, Ribble valley, Preston** |

This sequencing of alluvial deposits has a simple interpretation. Coarser sediments represent old channel deposits such as point and mid-channel bars. As stream channels migrate across floodplains, alluvial bars are abandoned and become widespread (Figure 5.15). Eventually, these coarse deposits are covered by fine sediments from flood events. The finer sediments, or overbank deposits, are the wash load that settles out of suspension as floodwaters spread across the valley floor.

Deposition of fine sediment on the floodplain surface results from the sudden reduction in flow velocity at the flood stage, as floodwaters spill out of the stream channel. The floodplain is much rougher than the channel and offers greater resistance to flow. As velocities slow, the deposition of suspended silt and clay takes place (Figure 5.20).

Rates of overbank deposition decrease away from the main channel. Closest to the channel, depositional rates may reach 3–5 mm a year. However, on average, floodplain surfaces accrete at much slower rates — around 0.25 mm a year. The exceptions are cut-off and abandoned floodplain channels. Here, deposition rates may be as high as 5–15 mm a year.

Figure 5.20 **Overbank deposits: fine sediments from the washload deposited by a recent flood event, North Burn, Teesside**

High-energy floodplains

High-energy floodplains develop in association with braided streams and rivers (Figure 5.21). Compared with low-energy floodplains they have:

- steeper gradients
- coarser sediments
- higher rates of erosion

High-energy floodplains are dynamic landforms because:

- braided channels constantly shift position across the width of the floodplain
- new alluvial bars develop continually
- there is little vegetation cover to anchor the coarse sediments

These sediments often have their source in debris cones and glacial moraines.

Figure 5.21 **High-energy floodplain, southern Iceland**

River terraces

Erosion on floodplains is mainly lateral. This causes stream channels to shift position as meanders migrate downstream and across floodplains. At the same time, gradual changes in the energy environment because of falling base levels or increases in discharge produce some downcutting. One effect of downcutting is the formation of new floodplains inside old ones. Remains of old floodplain surfaces often survive as terraces along valley margins. Unlike the terraces formed by rejuvenation, these terraces are at different levels or are unpaired (Figure 5.22).

The modern floodplain occupies only a small part of the valley floor, most of which consists of unpaired river terraces (T_1 and T_2). These remnant floodplains are made of coarse alluvial fill, which, in places, is being eroded and re-worked. The older terrace in Figure 5.22b (T_1) is approximately 3 m above the present-day flood level; the newer terrace (T_2) is 1–2 m above the flood level. Both terraces provide attractive, dry sites for villages, such as Arncliffe and Hawkswick.

Figure 5.22 **Unpaired terraces**

(a) Littondale, North Yorkshire

Unpaired river terraces (T_1 and T_2) at Littondale, North Yorkshire (FP = active floodplain)

(b) River Wharfe at Arthington, West Yorkshire

River terrace

Active floodplain

The sediment storage zone

Sediments enter long-term storage in lowland courses of streams and rivers. Sediment storage is responsible for landforms such as floodplains, deltas and estuaries.

Landforms in the sediment storage zone

Deltas

Deltas are areas of river-deposited sediment located at the mouths of rivers. Indeed, deltas occupy the mouths of many of the world's largest rivers, including the Amazon, the Ganges-Brahmaputra and the Mississippi. Some deltas project seawards, extending beyond the adjacent coastline. This process of building-out into the sea is called **progradation.** Progradation usually occurs where tide and wave energies are low.

Factors influencing the formation of deltas

Deltas form where rivers deposit sediment faster than wave action and tidal scour can remove it. Several conditions favour delta formation, including:

- rivers with very large suspended sediment loads, for example the Huang Ho in China, which transports 80 g of suspended sediment per litre of water. Rivers such as the Mississippi that drain large basins also deliver huge volumes of sediment to their mouths.
- a broad continental shelf margin at the river mouth that provides a platform for the accumulation of sediment. Deltas tend not to develop on narrow continental margins, such as the Pacific coast of America.
- rivers that flow into low-energy receiving basins (i.e. seas and lakes). These environments, with low-energy waves, have limited capacity to transport sediment deposited by rivers in the coastal zone.
- rivers that flow into receiving basins with a low tidal range. The absence of tidal scour favours the accumulation of sediment.

Delta morphology and processes

The main morphological feature of deltas is a **delta plain** (comparable to a floodplain) comprising river-deposited sediment. The **upper delta plain** (Figure 5.23) is furthest from the coast and is above the level of the highest tides. In contrast, the **lower delta plain** is submerged at regular intervals by high tides as well as river floods. Sediments accumulate in distributary bays forming mudflats and, eventually, salt marshes or mangroves. Accretion of fine sediments gradually fills these bays and elevates them above the high-tide level.

Figure
5.23
Delta structure

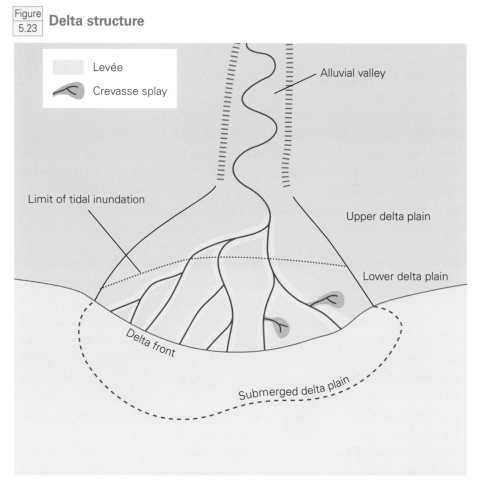

Distributary channels criss-cross deltas, splitting, merging and re-joining. They are bounded by natural embankments (**levées**) that form by deposition along channel margins. At high flow, levées may be breached. Water spilling out of distributary channels into adjacent basins experiences a sudden decrease in energy, causing rapid deposition. In these circumstances, **crevasse splays** may form. Essentially, a crevasse splay is a sediment lobe or small alluvial fan (Figure 5.24).

Extension of the lower delta plain seawards occurs at the mouths of distributary channels where levées are poorly developed. Here the flow is no longer confined. As a result, the flow expands, loses energy and builds up sandy distributary-mouth bars. Deposition also occurs as fresh water from the river meets saline water. This causes the tiny suspended sediment particles to stick together (**flocculation**) and settle out of suspension. As the distributary mouth bar grows, it splits the flow, causing the channel to bifurcate. Channels that converge or suddenly switch direction (**avulsion**) may merge.

The formation of distributary channels is accompanied by the development of levées. Levées often isolate areas of the lower delta plain between distributaries. Continual flooding by tides and rivers leads to gradual infilling and allows the growth of the delta seawards.

Figure 5.24 **Levées and crevasse splays in a delta distributary**

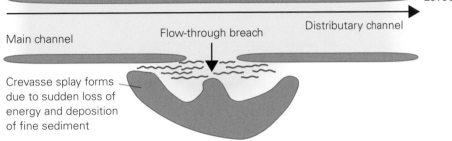
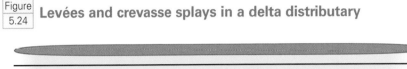

Levée

Main channel · Flow-through breach · Distributary channel

Crevasse splay forms due to sudden loss of energy and deposition of fine sediment

Types of delta

There are three main types of delta (Figure 5.25):

- river-dominant
- tide-dominant
- wave-dominant

The formation of distinctive delta types depends on:

- discharge
- sediment load
- waves and littoral currents, which redistribute sediment parallel to the coastline tides; tidal currents carry sediment perpendicular to the coast
- shelf width and slope, which influence how readily sediment is lost to the floor of the ocean basin. For example, a broad shallow delta such the Niger in west Africa could only develop on a broad, stable shelf.
- tectonics — a subsiding or stable platform. For example, a subsiding basin is in part responsible for the elongated bird's foot delta of the Mississippi.

The mouth of the Mississippi provides a good illustration of a **river-dominant delta** (Figure 5.26).

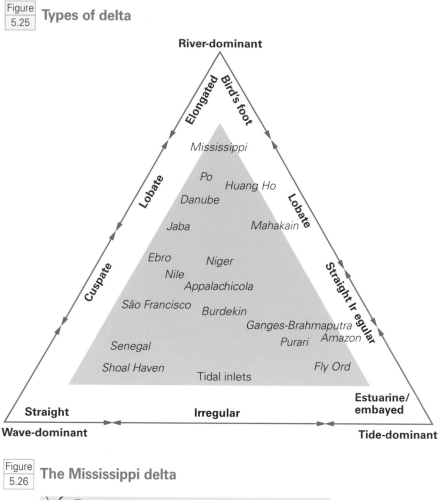

Figure 5.25 Types of delta

River-dominant

Elongated Bird's foot

Mississippi

Po

Huang Ho

Danube

Lobate

Jaba

Mahakain

Ebro

Niger

Nile

Appalachicola

Cuspate

São Francisco

Burdekin

Ganges-Brahmaputra

Amazon

Senegal

Purari

Straight Ir egular

Shoal Haven

Fly Ord

Tidal inlets

Lobate

Straight

Irregular

Estuarine/embayed

Wave-dominant

Tide-dominant

Figure 5.26 The Mississippi delta

0   km   100

N

The river flows into the low-energy Gulf of Mexico, where there is a small tidal range and low-energy wave action. It drains a huge basin and carries a large sediment load (Figure 5.27).

Figure 5.27 **The Ganges–Brahmaputra delta**

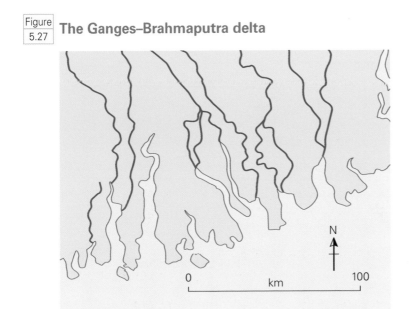

Tide-dominant deltas resemble estuaries. Examples include the Ganges–Brahmaputra and the Rhine.

A high tidal range leads to strong tidal currents. With moderate wave energy, longshore currents are weak and sediments are deposited perpendicularly to the coastline. Tide-dominated deltas are inundated at high tide and drained through distributary channels and creeks at low tide. The greater the tidal range, the more the seaward development of the delta is constrained.

Wave-dominant deltas have smooth coastlines, well-developed beaches and dune systems. Strong longshore currents move sediments along the coast, removing fines, and forming bars parallel to the coastline. Transverse bars give rise to an interrupted barrier coastline, and dam tidal lagoons to their rear.

The human impact on deltas

In drainage basins such as the Amazon, where deforestation has caused accelerated soil erosion, increased sediment loads have led to rapid progradation and delta growth. However, it is more usual for human impact to cause delta shrinkage. Dam building and a reduction in sediment delivered to deltas is often to blame. For example, the Nile delta (Figure 5.28) has experienced significant coastal erosion since the completion of the Aswan High Dam in 1970. Even more dramatic is the impact of human activity on the Colorado delta. So much water has been diverted from the Colorado River that, today, virtually no river water and sediment enters the delta.

Figure 5.28 **The Nile delta**

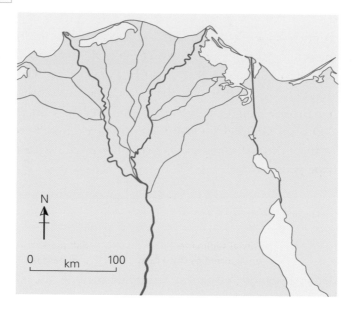

Table 5.4 **River deltas: discharge, sediment load, tidal range and wave energy**

River	Basin area (km$^2 \times 10^3$)	Discharge mean (cumecs)	Sediment (tonnes yr$^{-1} \times 10^6$)	Tidal range (m)	Wave power (ergs $\times 10^7$)
Amazon	5888	149 726	1200	4.9	0.193
São Francisco	602	3420	6	1.86	30.42
Nile	2716	1480	50	0.5	10.25
Niger	1113	8769	5	1.4	2.01
Mississippi	3344	15 631	495	0.5	0.034
Huang Ho	865	2571	2080	1.13	0.218
Ganges–Brahmaputra	1597	34 500	1670	3.6	0.586
Irrawaddy	342	12 558	285	2.7	0.193
Danube	713	6250	80	0.09	0.034
Indus	888	4274	480	2.6	14.15

Estuaries

Estuaries are the funnel-shaped, tidal mouths of rivers (Figure 5.29). Most estuaries around the British Isles are essentially drowned river valleys, flooded by rising sea level in the period 20 000 to 6000 years before present. Some estuaries occupy broad lowland valleys (e.g. the Humber estuary); others have developed in deep, narrow valleys along upland coasts (e.g. the Dart estuary). During the past 6000 years, sea level has stabilised. Infilling through sedimentation and the growth of mudflats and salt marshes has given estuaries their smooth, funnel-shape plan.

Figure 5.29 **The Kent estuary, Cumbria**

Older infill and land reclamation — Fluvial sediments transported by River Kent — Salt marsh and recent sedimentation — River Kent (shifting channel) — Sand flats exposed at low tide — Flood tide sediments from Morecambe Bay

Estuaries are sediment traps. More sediment is carried into estuaries by rivers and tidal currents than is carried out — hence the development of mudflats and salt marshes. Deposition of silt and mud occurs through fall-out from suspension by gravity and from flocculation. The presence of estuaries (and the complete absence of deltas) around the coastline of the British Isles is explained by:

- exceptionally strong tidal scour caused by a large tidal range. In estuaries such as the Severn and the Kent the spring tidal range exceeds 10 m (Figure 5.30). Strong tidal scour prevents the growth of channel bars and channel bifurcation that might otherwise lead to the formation of deltas.
- rivers that drain relatively small catchments and therefore have modest sediment loads
- river catchments that are well vegetated and therefore have relatively low sediment yields

A **bore** is an extreme example of the flood tide that occurs on estuaries with a large tidal range. The flood tide scours the river channel and is responsible for the net sediment gain in the estuary (see Figure 5.30).

| Figure 5.30 | **Tidal bore in the Kent estuary, Cumbria** |

Environmental change and fluvial landforms

We saw in Chapter 1 that streams are energy systems that move gradually towards equilibrium. Equilibrium is reached when streams are able to transport their sediment load with minimal energy expenditure. Equilibrium is an average condition and it can only be maintained for as long as environmental conditions remain stable. However, environmental conditions are rarely stable for long periods of time. For example, in the past 15 000 years, frequent changes in climate, discharge, tectonic movement and sediment load have disrupted stream equilibrium in the British Isles. Streams have responded to environmental changes by adjusting the processes of degradation and aggradation. As a result, new and distinctive landforms have been created.

Climate change

Climate change disrupts stream equilibrium on a continental scale. Fifteen thousand years ago, the Earth's climate warmed as the Devensian glacial period drew to a close. Ice sheets and glaciers, which had covered most of northern Europe for the previous 100 000 years, slowly retreated, leaving behind huge amounts of rock debris. In the immediate postglacial period, cold climatic conditions supported only sparse vegetation cover. Meanwhile, streams swollen by summer meltwater had higher discharges than today. Under these conditions, streams had unusually high energy, high sediment loads and braided channels. As a result, aggradation was widespread, and valley floors were rapidly in-filled with coarse alluvium.

Climate change affects sea-level change too. During glacial periods, sea level falls by an average of 100 m. With lower base levels, streams have surplus energy and begin to cut down towards their new base levels. In the process, new landforms are created.

Stream erosion, renewed by either a fall in base level or higher discharge, carves a new floodplain within the old one. The remnants of the original floodplain survive as **river terraces**. Where downcutting is particularly rapid, these terraces are often 'paired' at the same level (Figure 5.31). Paired terraces are distinct from 'unpaired' terraces. The latter form as streams migrate across their floodplains and at the same time gradually degrade their beds.

Figure 5.31 **Development of paired river terraces**

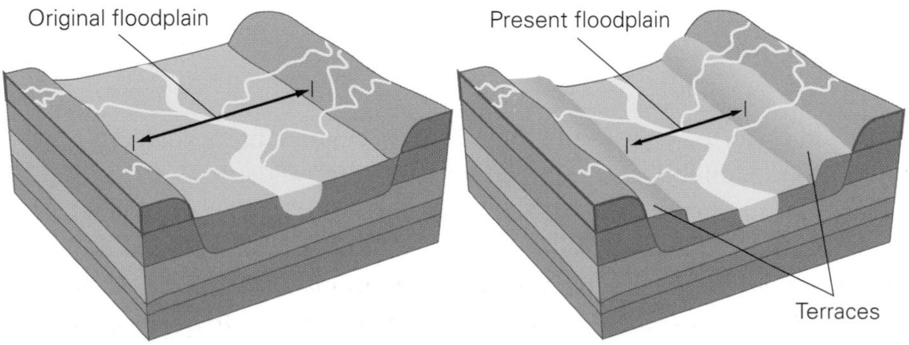

Original floodplain | Present floodplain

Terraces

Tectonic uplift

Pressure, tension and volcanic activity within the Earth's crust often result in regional tectonic movement. Movements such as tectonic uplift occur on geological time scales and can have profound effects on streams. Higher elevations and steeper gradients provide streams with additional energy inputs. Surplus stream energy often initiates a prolonged period of erosion that produces spectacular landforms such as **canyons**, **incised meanders** and **knickpoints**.

The Colorado plateau

The Colorado plateau in the southwest USA covers much of Utah, Arizona and western Colorado (Figure 5.32). The plateau is drained by the Colorado River and its main tributaries, the Green River, the Little Colorado River and the San Juan River. Twenty million years ago, the Colorado plateau was a lowland region and it was then that the present-day drainage system was established.

Figure
5.32
Map of the Colorado plateau and drainage system

Despite a vertical uplift of almost 3 km, the main streams crossing the Colorado plateau have maintained their original courses. The extra energy from uplift rejuvenated streams that cut ever-deeper valleys. The best known of these streams is the Colorado River. It carved the 2 km-deep Grand Canyon in less than 6 million years.

Meanwhile, tributary streams such as the San Juan River also eroded vertically into the sedimentary rock. The dry climate, solid rock walls and limited slope processes caused minimal back-wasting of valley slopes. As a result, stream channels were deeply incised, leading to the formation of **entrenched meanders** on the San Juan River (Figure 5.33)

Figure
5.33
Entrenched meanders on the San Juan River, Colorado plateau, Utah

Throughout the Colorado plateau, tributary streams experienced falls in base levels as the major rivers cut down. They responded by eroding new long profiles as knickpoints migrated upstream. For example, Bright Angel Canyon, Arizona (Figure 5.34), was formed by rapid vertical erosion of the Grand Canyon by the Colorado River, which led to a fall in the base levels of tributary streams and subsequent rejuvenation.

Figure
5.34
Bright Angel Canyon, Arizona

Human activities

Human activities can also increase stream sediment load. In the British Isles, deforestation, linked to the introduction of agriculture 5000 years ago, caused a rapid increase in soil erosion. Eroded soils, transported by surface wash and overland flow, would have entered stream systems and increased suspended loads. The thick layers of silty deposits found at the top of the alluvial sequence in most floodplains in Britain today are evidence for this process.

6 Human influences on river flow

Human activities modify river flows to varying degrees. These modifications may be direct or indirect (Figure 6.1). In this chapter we shall concentrate on how some human activities can affect river flows indirectly.

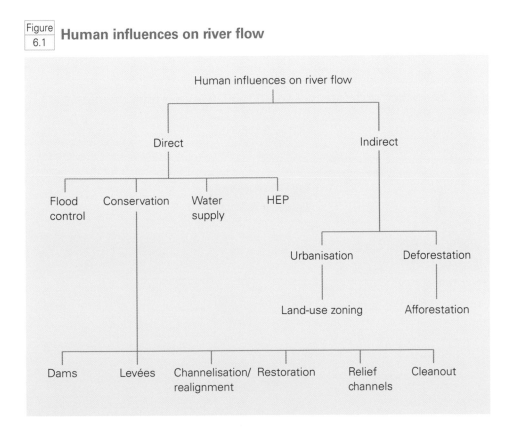

Figure 6.1 **Human influences on river flow**

Land-use changes such as deforestation and urbanisation often bring about unintended modifications to river flow. On the one hand, deforestation and urbanisation reduce water storage and base flow; on the other hand, they increase runoff and storm-water flow. The implications of change are stark — more frequent peak flows and increased risks of flooding.

Deforestation

Draix, southeast France

The Laval and Brusquet catchments at Draix in southeast France illustrate the effects of deforestation on stream flow (Figure 6.2). Both catchments have similar precipitation (800–900 mm per year), geology (black marl) and relief (700–1200 m). However, they are very different in one respect — forest cover. The Laval basin has been extensively deforested — today just 22% is forest. The neighbouring Brusquet catchment still has 87% forest cover.

Forested catchments both reduce and delay runoff. Precipitation is intercepted by forest trees, and this delays the movement of water to the ground. Meanwhile, evaporation of intercepted rainwater from leaf surfaces and transpiration results in moisture loss and reduces runoff (Figure 6.2).

Figure 6.2 **The effect of deforestation on rainfall–runoff relationships**

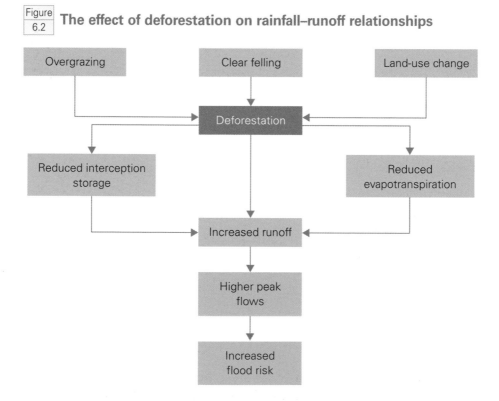

Peak flows in the deforested Laval catchment are four to five times higher than in the Brusquet catchment (Figure 6.3). This increases greatly flood risks on the

Laval. Flood risks are also raised by high sediment loads in the Laval catchment (177 tonnes ha^{-1} compared with 4 tonnes ha^{-1} in the Brusquet catchment). These high sediment loads are themselves a consequence of deforestation. They increase rates of channel deposition and reduce channel capacity.

Activity 1

Study the hydrographs in Figure 6.3 that show precipitation and discharge over a 48-hour period for the Laval and Brusquet rivers.

Figure 6.3 **Floodwater in the Laval and Brusquet watersheds, 8 and 9 March 1991**

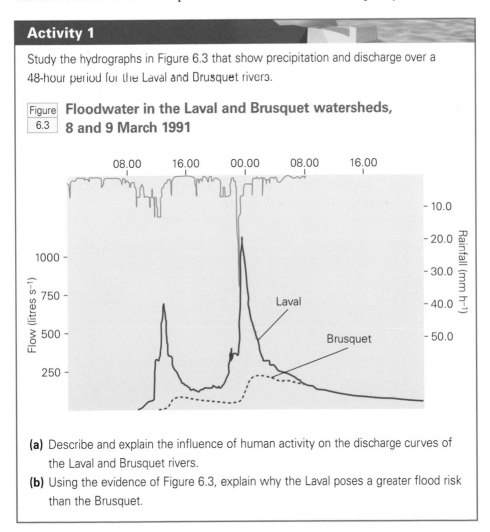

(a) Describe and explain the influence of human activity on the discharge curves of the Laval and Brusquet rivers.

(b) Using the evidence of Figure 6.3, explain why the Laval poses a greater flood risk than the Brusquet.

Urbanisation

Urbanisation is the conversion of land use from rural to urban. Farmland, grassland and woodland are replaced by housing, offices, factories and roads. Artificial urban surfaces have a very different hydrology from natural surfaces covered by vegetation and soil (Table 6.1 and Figure 6.4).

Figure 6.4 **The impact of urban development on runoff**

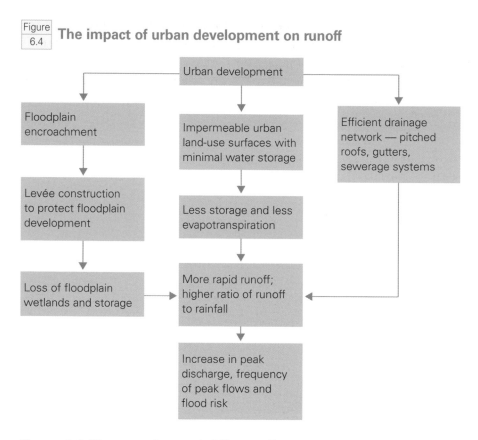

Figure 6.5 illustrates how rainfall–runoff relationships in urban areas are modified by impermeable surfaces and artificial drainage networks.

Figure 6.5 **Heavy rain in an urban area**

- Urban areas have many impermeable surfaces such as tarmac, brick, tiles and concrete, which provide minimal storage capacity and buffers to rapid runoff.
- Urban areas have elaborate drainage networks — pitched roofs, gutters, sewerage systems — which remove water quickly and efficiently. The result is short lag times and high, but short-lived, peak flows.

Activity 2

Describe and explain the main features of the hydrographs in Figure 6.6:
- before urbanisation
- after urbanisation

Figure
6.6

Hydrographs for a catchment before and after urbanisation

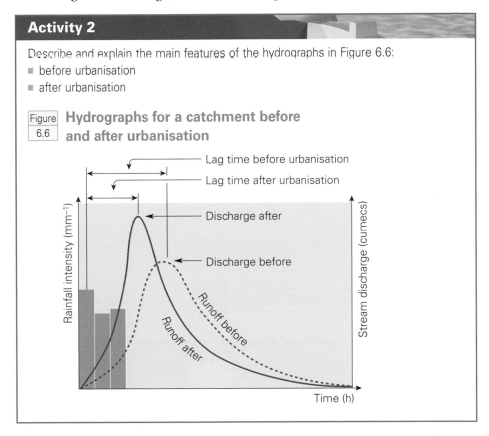

Apart from the transformation of land use, urbanisation also encroaches on floodplains, greatly increasing the exposure of urban populations to flooding. In Japan, three-quarters of the country's population and properties are located on floodplains. In the UK, 2 million homes and 185 000 businesses sited on floodplains are at risk. The cost of flooding in the UK in 2000 topped £1 billion.

There is a second way in which urban encroachment on floodplains increases flood hazards. Normally the wetlands on floodplains are natural storage areas for floodwaters. Inevitably, urbanisation results in drainage and loss of these valuable storage areas. Meanwhile, rivers, boxed-in by levées, are no longer able to flood. Therefore, more water gets into rivers more quickly, and less is stored. Little wonder that peak flows increase in magnitude and frequency, as do the risks of flooding downstream.

Urbanisation in the Yasu drainage basin (Honshu)

The Yasu River drains a small catchment of 400 km² around Lake Biwa in the Shiga Prefecture in southwest Japan. Between 1975 and 1997, urban land use in the catchment increased from 4.6% to 6.9%. Urbanisation has modified rainfall–runoff relationships in the Yasu catchment. There has been a noticeable increase in flood peaks (Figure 6.7) and, since 1985, storm hydrographs (Figure 6.8) have had sharper profiles, with higher peaks, and steeper rising and recessional limbs. These changes reflect the reduced storage of water in urban areas caused by their impermeable surfaces, the loss of storage space and more efficient drainage systems.

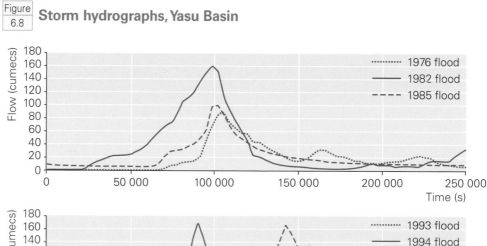

Figure 6.7 **Flood peaks in the Yasu basin, 1975–2000**

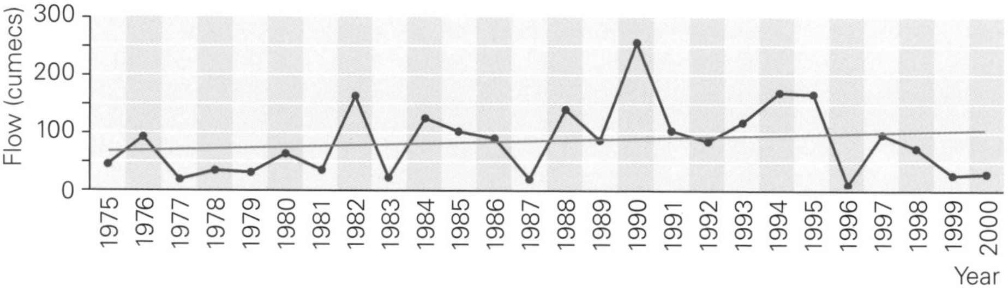

Figure 6.8 **Storm hydrographs, Yasu Basin**

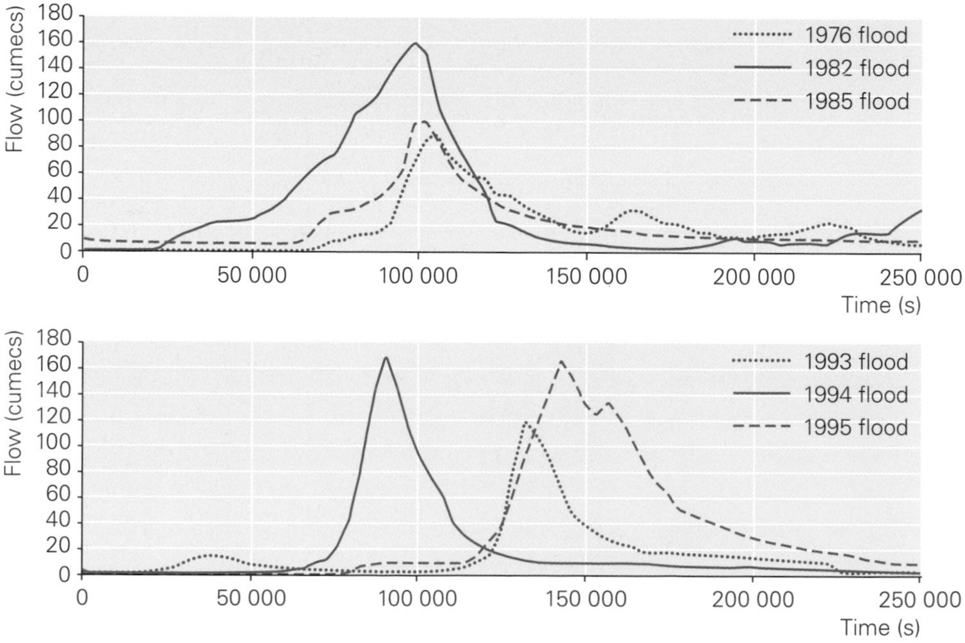

Activity 3

The rainfall coefficient is the ratio of the amount of water not absorbed by a surface to the total amount of rainfall during an event.

Table 6.1 **Rainfall coefficients for urban areas**

Land use	Surfaces	Rainfall coefficient
Commercial	CBD	0.70–0.95
Residential	Single family detached	0.30–0.50
	High-density terrace	0.60–0.75
Industrial	Light	0.50–0.80
	Heavy	0.60–0.90
Streets	Tarmac	0.70–0.95
	Concrete	0.80–0.95
	Brick	0.70–0.85
Roofs	Tile/slate	0.75–0.95
Lawns	Sand-soil flat	0.05–0.10
	Heavy clay steep	0.25–0.35

Study the rainfall coefficients for urban areas given in Table 6.1.

Suggest reasons for the rainfall coefficient values in:

(a) the CBD

(b) residential areas

(c) areas of light and heavy industry

7 Approaches to river management

Structural approaches

River flow and fluvial processes may be deliberately modified in order to:

- develop the resource potential of a river
- reduce river-related hazards, such as flooding and erosion
- restore the ecological and geomorphological balance to rivers degraded by human activity

Direct action is often justified in the name of resource development. Examples include:

- dam construction to provide hydroelectric power and water supplies
- improvements to navigation through dredging and channelisation
- developing recreational attractions, such as reservoirs

The abstraction of water from rivers is commonplace. Although not a deliberate attempt to change flow conditions, abstraction alters the energy environment of a river and its ecology.

It is important to remember that rivers are energy systems and that human action that changes flow patterns and fluvial processes will ultimately impact elsewhere in the system. Such impacts are sometimes unexpected, and often environmentally damaging.

Flow control can be achieved by structural and non-structural measures:

- Structural measures cover hard engineering works such as dams, levées, channel straightening, clean-out (removing debris from the channel), sluice gates and relief channels (Figure 7.1 and Table 7.1).
- Non-structural methods usually focus on land-use change, especially afforestation in headwater catchments and the zonation of land on floodplains.

<table>
Figure
7.1
</table>

Structural controls

(a) Dam and reservoir — Scar House and reservoir, Nidderdale

(b) Bank reinforcement — armour blocks to protect the bank from erosion

(c) Willow spiling — stabilises the bank as the willow starts to grow

(d) Gabions — wire mesh cages filled with small rocks — provide hardened bank protection

Table 7.1 Some protection methods against streambank erosion

Method	Characteristics
Riprap — large angular stones and boulders used on stream banks where velocities are too high to establish vegetation	▪ Effective in stopping bank undercutting ▪ Must be large enough to resist movement and transport downstream
Gabions — wire mesh cages filled with small rocks, providing hardened streambank protection	▪ Provide long-term stabilisation, as long as the bed–gabion interface is protected with rock riprap ▪ A common substitute for riprap where only small stones are available
Vegetation cover — often used on small tributaries	▪ Has the advantage of being self-repairing ▪ Roots bind the bank material together ▪ Stems, branches and foliage provide resistance to flow and absorb energy
Willow spiling	▪ A 'living' revetment ▪ Live willow stakes are interwoven with thin willow branches to create wattle walls ▪ The growing willow stabilises the banks
Wing dykes — obstructions to flow that protrude into the stream at right angles to the bank	▪ May be steel, concrete, wood or rock structures ▪ Deflect flow away from eroding banks ▪ Reduce flow velocity and cause sedimentation, which protects against bank erosion

Flow regulation by dams and reservoirs

The construction of dams and reservoirs has both upstream and downstream effects on stream channels. Upstream, the effects include delta formation, a gradual raising of stream levels and more pronounced meandering. The downstream effects are often greater. They include the entrapment of sediment in reservoirs and changes in discharge. Overall, the impact of reservoirs on flow regimes is to reduce peak flows and increase low flows.

Reservoirs capture all incoming bedload and a high proportion of suspended load. With minimal sediment load, a river exiting a reservoir has excess energy and immediately starts to degrade the channel profile below the dam. However, erosion and degradation are not inevitable. If bedload inputs from tributaries downstream of the reservoir are high, the downstream channel may in some circumstances even aggrade. In these conditions, reduced flood peaks are unable to transport the sediment inputs.

Little Tallahatchie River: dam construction and feedback responses

The Little Tallahatchie River is a meandering sand-bed tributary to the Yazoo River in north-central Mississippi (Figure 7.2). About 85% of its basin is controlled by the Sardis dam. This dam, built for flood control in 1939, is located about 36 km above the confluence of the Little Tallahatchie with the Yazoo. In the post-dam period, some channel modifications (e.g. clearing, snagging and straightening) were carried out downstream of the dam.

Figure 7.2 Flood control dams on tributaries of the Yazoo River, Mississippi

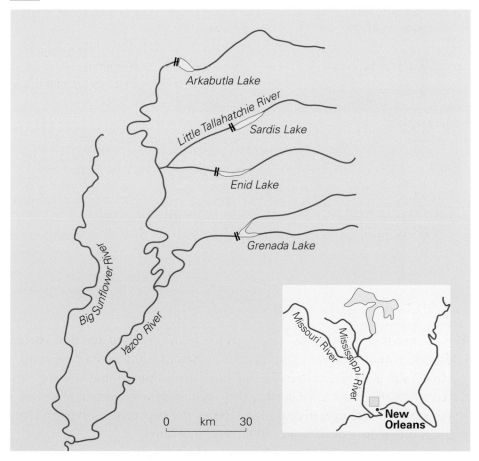

The Sardis dam had a massive effect on river discharge — whereas pre-dam floods had exceeded 566 cumecs, post-dam floods reached a maximum of just 184 cumecs.

Sediment entrapment behind the Sardis dam triggered **clearwater erosion** and the progressive flattening of the channel slope below the dam. The combined effects of bed degradation and smaller flood discharges led to a reduction in water levels at the mouths of tributary streams. Tributary streams were rejuvenated by this fall in base level. As a result, bed degradation progressed upstream where it caused bank collapse, meandering and lateral erosion that delivered large quantities of coarse sand and gravel to the Little Tallahatchie. The reduced flows of the Little Tallahatchie were unable to transport all this material. Therefore, as a secondary response, the Little Tallahatchie channel (except for the 5 km section immediately below the dam) started to aggrade.

The complex response of the Little Tallahatchie to the Sardis dam has required dredging to maintain flood capacity and flood control. Erosion protection and grade controls have been required on several tributaries to reduce loss of land and delivery of coarse sediment.

This example shows how initial and final instability responses — degradation followed by aggradation — can operate in opposite directions.

Activity 1

(a) Study Figure 7.3. Describe and explain the differences in the flow–duration curves of the Little Tallahatchie before and after construction of the Sardis dam.

(b) Draw a flow diagram to summarise the complex downstream response of the Little Tallahatchie River and its tributaries to the building of the Sardis dam.

Figure 7.3 **Effect of a storage reservoir on the Little Tallahatchie downstream flow–duration curve**

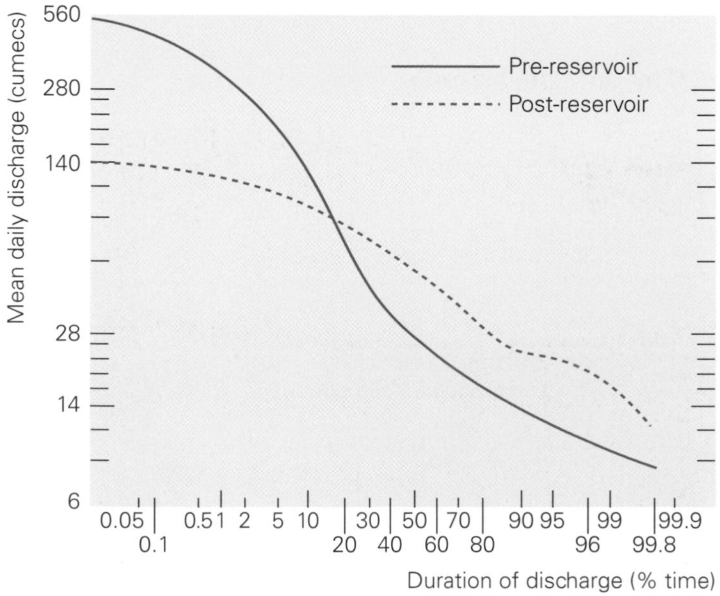

Activity 2

Study Figure 7.4. Explain the upstream response of the river to dam construction.

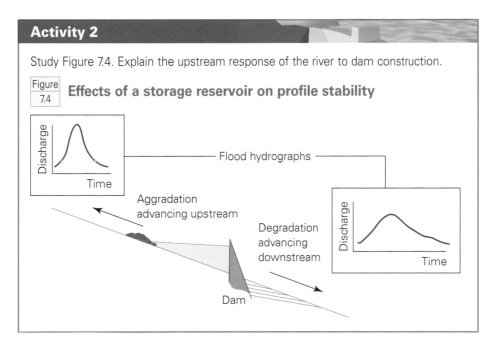

Figure 7.4 **Effects of a storage reservoir on profile stability**

Glen Canyon dam and Lake Powell

The Glen Canyon dam (Figures 7.5 and 7.6) in the Lower Colorado basin was completed in 1963. The dam created a huge reservoir — Lake Powell — over 300 km long. Lake Powell took 18 years to fill and in the process vast areas of canyon lands in Utah were flooded.

Figure 7.5 **Glen Canyon dam, Arizona**

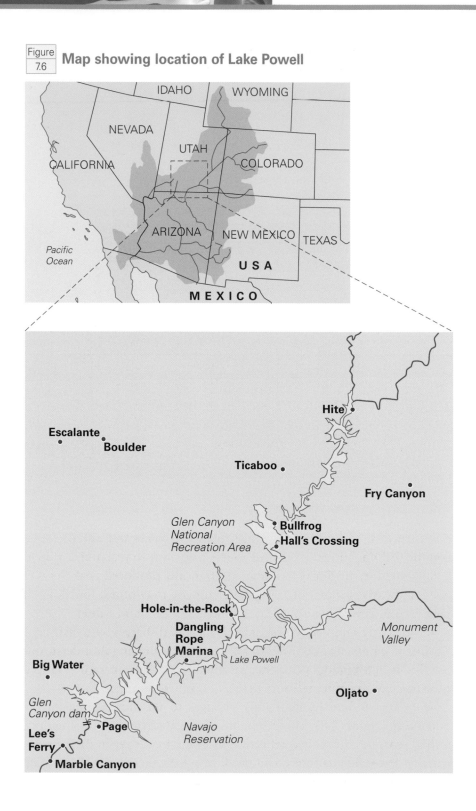

Figure 7.6 **Map showing location of Lake Powell**

The Glen Canyon dam is a multipurpose scheme:

- It generates over 4000 MW of hydroelectric power.
- It provides 10 billion m³ of water a year to California, Arizona, Nevada and Mexico (mostly for irrigation).
- Lake Powell is a major tourist attraction, especially for water-based recreational activities, such as fishing, boating and water sports (Figure 7.7).

Figure 7.7 **Lake Powell, the multipurpose reservoir created by the Glen Canyon dam**

Impact of the Glen Canyon dam on the Colorado River

Before construction of the dam, flow rates, sediment loads and water temperatures on the Colorado River fluctuated widely from year to year and season to season. Snowmelt in the Rocky Mountains commonly produced peak flows in late spring and early summer of 2500–3000 cumecs. In contrast, low flows of less than 85 cumecs were typical of late summer, autumn and winter.

In spring and early summer, sediment loads increased sharply with discharge. Large sediment inputs also occurred in late summer following thunderstorms and flash floods in tributary catchments. Meanwhile, water temperatures ranged from near freezing in winter to more than 27 °C in late summer.

The Glen Canyon dam brought huge changes to the flow regime of the Colorado River (Figure 7.8). Flows became dominated by daily fluctuations (rather than seasonal variations) reflecting short-term changes in the demand for electricity. These fluctuations could produce a difference in river level of up to 9 m in any given day.

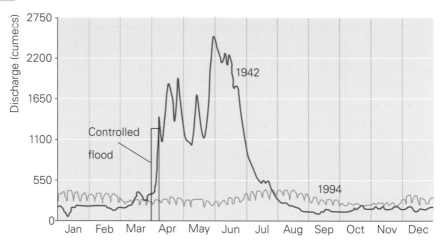

Figure 7.8

Annual flow on the Colorado River, 1942 and 1994

Following the building of the dam, mean daily flows exceeded 850 cumecs only 3% of the time (compared with 18% before dam construction). On the other hand, low flows became less frequent. Today, flows of less than 140 cumecs occur around 10% of the time (compared with 16% before dam construction).

The dam greatly reduced the sediment load of the Colorado River. All sediments from upstream of the dam are now trapped in Lake Powell. Thus any downstream sediment comes entirely from tributary streams. In contrast to the pre-dam situation, water for electricity generation, drawn from deep below the surface of Lake Powell, has caused a dramatic fall in river temperatures to just 8°C. All of these changes have had a major impact on the geomorphology and ecology of the Colorado River.

Impact of changes in flow on the geomorphology and ecology of the Colorado River

Before 1963, snowmelt in the upper Colorado River basin produced floods averaging 2800 cumecs. These floods scoured large volumes of sediment from the river channel bed. Receding floodwaters then deposited the sediments along the river banks as large sand bars. This annual process of scour and fill not only maintained channel sand bars but also kept them clear of vegetation. The virtual elimination of floods in the post-dam period, together with large daily fluctuations in flow and sediment entrapment in Lake Powell, has led to the erosion of many sand bars. At the same time, debris fans, formed where tributaries from side canyons bring in large amounts of coarse sediment, have built up in the main channel.

The post-dam flow regime has also had adverse ecological effects. Without the scouring action of major floods, exotic vegetation such as tamarisk has colonised large parts of the river, pushing out native species such as cottonwood. Indigenous fish species, which evolved in the warm turbid waters of the Colorado, have been hard hit too. Three of the eight native fish species have disappeared and two others — the razorback and humpback chub — are endangered. The decline in native fish species is also related to the loss of backwater lagoons that in the past acted as nursery areas for young fish, and to competition from exotic species such as trout and catfish, which thrive in cold, clear water.

Management response: controlled flooding

In an attempt to restore balance to channel landforms and the ecology of the Colorado River, an artificial flood was created on the river between 26 March and 2 April 1996. This was the first time hydrologists had ever manufactured a large-scale flood —1275 cumecs were released from the Glen Canyon dam. The total cost of the experiment was around $4 million, mainly due to lost electricity production.

The flood peak simulated spring flows before the construction of the Glen Canyon dam. The main aim was to rebuild sand bars in the river channel. It was argued that the flood would have sufficient energy to entrain sand from the channel bed and deposit it at higher elevations on the river banks. At the same time the flood would scour backwater reaches, remove fine sediment and vegetation and clear debris fans.

It did succeed in forming many high-elevation sand bars and removed sand from the river channel, transferring it to long-term storage on the banks. However, by 1999 sand storage had again fallen to pre-flood levels and low-level sandbanks in the main channel had re-formed. Further floods are planned in future.

Activity 3

Draw a flow diagram to show how the Glen Canyon dam and Lake Powell have affected the Colorado River system.

Levées

The purpose of levées or flood embankments is to confine peak flows within stream channels and thus protect floodplain developments from inundation (Figure 7.9). Occasionally, levées are set back from stream channels to allow a limited area of the valley floor to flood.

Figure 7.9 **Levées on the River Wyre**

However, levées are controversial. Unless they are set back, levées prevent flood waters from spilling onto adjacent floodplains. Thus, by preventing flooding, they tend to increase peak flows. Furthermore, rivers contained by levées may flow several metres above adjacent floodplains, thus posing an even greater risk to settlements if they are breached.

Hard-engineered flood-control structures on the Mississippi River

Over the past 70 years, many hard-engineered flood-control structures have been built along the Mississippi River and its tributaries. These structures, mainly developed by the US Army Corps of Engineers, were built to control flooding and to maintain navigation along the river.

The most impressive of these control structures is a 12 m high levée system. As well as being confined by levées, the Mississippi has been straightened by artificial meander cut-offs to increase flow velocities and speed the transfer of flood peaks. Navigation structures include a 2.7 m navigable channel, supported by riprap (rock armour) and revetments to stabilise the position of the channel, and wing-dykes. These act like groynes in coastal environments. They trap sediment, narrow the channel and thus increase flow velocities and scour. This reduces channel aggradation and maintains sufficient water depth for navigation. In addition, 28 locks and dams have been built on the upper Mississippi to guarantee 2.7 m water depth in times of low flow, permitting ships to move over knickpoints.

The Mississippi floods of 1993

In 1993, the upper Mississippi basin experienced its worst ever floods. Over 150 major rivers and their tributaries burst their banks. Unprecedented flood levels were measured at many stations in Illinois and Missouri. At St Louis, a new record height of 15.06 m (surpassing the old record by more than 1.5 m) was reached on 1 August and a record flow of 28 320 cumecs occurred (Figure 7.10). In all, the 1993 floods claimed 48 lives. Economic costs approached $20 billion and more than 50 000 homes were either destroyed or damaged, making the Mississippi floods the second most costly natural disaster in US history.

Figure 7.10

Mississippi River: hydrograph for St Louis, 1 July–10 August 1993

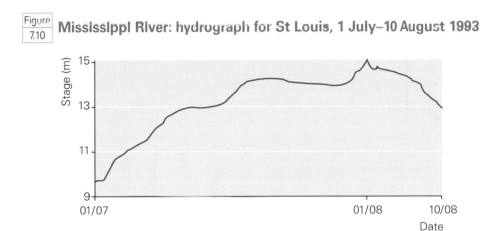

Most flood damage was caused by levée failure. The first levée was overtopped on 7 June; altogether, more than 1000 levées failed.

The effect of levées on the 1993 floods is controversial. The rainfall events that led to record river levels were exceptional, with an estimated recurrence interval of 75–300 years. However, many experts argued that flooding was made worse by the levées, which artificially raised flood peaks to record levels.

Levées not only appeared to increase the flood risk but, by raising flood heights and channel discharge, they also increased river energy levels. Much of this surplus energy was expended in erosion, widening the channel and lengthening meander bends.

It is now widely believed that long-established levée systems are responsible for the complete reforming of meanders, with slope flattening through degradation upstream and steepening by aggradation downstream. By limiting sediment deposition on floodplains, levées also accelerate deposition within stream channels, thus reducing channel capacity and increasing flood risks.

In actively meandering channels there is another danger — continued meander migration, aggravated by increased in-channel discharges, could encroach on levée set-back distances and attack the levées themselves. Where populations depend on levées for security, the risk posed during large floods may be extreme.

Clearing and snagging

Clearing and snagging is used when stream channels are restricted by dense vegetation, accumulations of drift debris or blockage by uprooted trees. Removing these obstructions reduces the flood risk by increasing the channel cross-sectional area and flow velocity. The procedure involves the removal of log jams, sediment blockages, large trees and underbrush. The floods on the River Derwent at Malton (North Yorkshire) in 1999 were blamed by some local residents on the failure of the Environment Agency to clear bankside vegetation.

Realignment or channelisation

Realignment has been used widely as a flood-control measure to increase channel capacity and reduce the loss of land caused by meander migration. Realignment sometimes involves the replacement of a meandering length of channel by a straight channel or the elimination of selected meander bends by cut-offs. Increased capacity results mainly from a steeper average gradient in the straightened channel (Figures 7.11 and 7.12).

Figure 7.11 **Channel straightening on the River Lune, Cumbria; the banks have been reinforced with armour blocks and gabions**

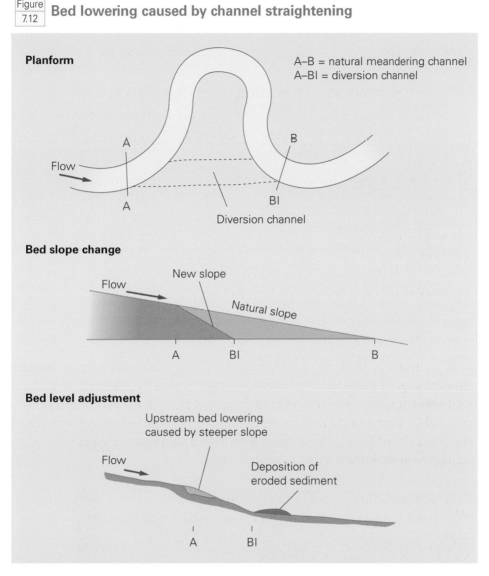

Figure 7.12 **Bed lowering caused by channel straightening**

The response of a stream to realignment is variable. In some examples, with gentle slopes and erosion-resistant channels, realignment does not cause serious problems, especially where flows are regulated by reservoirs. Elsewhere, channel straightening has led to channel degradation, bank erosion and tributary incision. In such cases, realignment must be accompanied by gradient control structures (e.g. weirs) to check velocity, and bank protection to prevent the development of new meanders. In straightened channels, bed materials scoured from upstream often increase sediment loads and cause problems through channel deposition downstream.

Flood-relief channels

Flood-relief channels reduce flood risks by diverting a proportion of peak flow and lowering water levels in the main channel.

Maidenhead, Windsor and Eton flood-relief channel

The towns of Maidenhead, Windsor and Eton and nearby villages have a long history of flooding by the River Thames. On average, the area floods about once every 5–7 years, though the last major flood was in 1947. It is estimated that a similar flood today would damage more than 5500 properties, affect over 12 500 people and cost £40 million. Major roads in this area, including the M4, would be severely interrupted.

In 1995, a flood-control scheme based on a flood-relief channel was approved (Figure 7.13). Alternative management schemes such as flood storage, levées and dredging to increase channel capacity were rejected as either impracticable or environmentally unacceptable. Completed in 2001 at a cost of £84 million, the flood relief channel (known as the Jubilee River) provides protection for a 1 in 65 years flood event. The Jubilee River leaves the Thames in North Maidenhead and rejoins the main river 11.8 km downstream at Windsor (Figure 7.13). It is 45 m wide and has an average depth of 5 m.

| Figure 7.13 | **The Jubilee River: flood-relief channel from the River Thames** |

In the Maidenhead–Windsor area, the River Thames has a bankfull flow of around 285 cumecs. The Jubilee River diverts excess water from the Thames, keeping its discharge below bankfull flow. The amount of water flowing in the Jubilee River is controlled by gates near Taplow Mill.

The Thames and the Jubilee River have a combined maximum discharge of 515 cumecs, with the Jubilee River carrying 215 cumecs. Under normal conditions the Jubilee River has running water all year round. This flow is derived naturally from groundwater and there is also a small flow of about 10 cumecs diverted from the Thames.

Performance of the Jubilee River during the January 2003 floods

The period between 1 November and 31 December 2002 was exceptionally wet. Rainfall along the Thames Valley was 261 mm — more than double the average. As a result, by early 2003 the Thames was close to record levels. Widespread flooding occurred throughout the Thames catchment. However, the Maidenhead–Windsor area, protected by the Jubilee River, remained flood-free. Peak flow in the Jubilee River reached 144 cumecs, while flow in the Thames was held below bankfull, at 260 cumecs. No properties were threatened, and the entire stretch of the river and floodplain from Maidenhead to Windsor remained on only a 'flood watch' warning (see Chapter 8). It is estimated that the flood-relief channel saved over 400 properties from flooding.

Several places downstream of the Jubilee River were not so lucky. Residents in these areas blamed the flood-relief channel (if the Thames had flooded in the Maidenhead–Windsor area, places downstream would have been saved). However, the Jubilee River did not increase the flood peak — it merely divided the flow between Maidenhead and Windsor. With discharge well in excess of bankfull capacity, it is highly likely that places downstream of Windsor, such as Datchet and Wraysbury, would have flooded anyway.

Despite their benefits, flood-relief channels have some disadvantages. They upset the delicate balance between width, depth and gradient in stream channels. This balance is finely adjusted to discharge and sediment load. Streams often respond by channel aggradation, reducing channel capacity. Additional problems include sourcing land for channel construction and devising systems that permit flow diversion only during floods.

Activity 5

Suggest how building a flood-relief channel, such as the Jubilee River, could cause changes to the natural channel of the River Thames.

Sluice gates and flood basins

Although sluice gates and flood basins are hard-engineered structures, they are more environmentally friendly than other hard structures. This is because they partly mimic natural processes.

When there is a serious risk of flooding, sluice gates in the stream bed are raised. These act as a dam or weir and divert the flow onto the adjacent floodplain. The floodplain acts as a temporary storage basin or washland for the floodwater. This reduces stream flows and prevents flooding downstream.

This type of scheme was adopted on the River Wyre at Garstang and Catterall in north Lancashire, following serious flooding in 1980 (Figures 7.14 and 7.15).

Figure 7.14 **Drainage basin of the River Wyre**

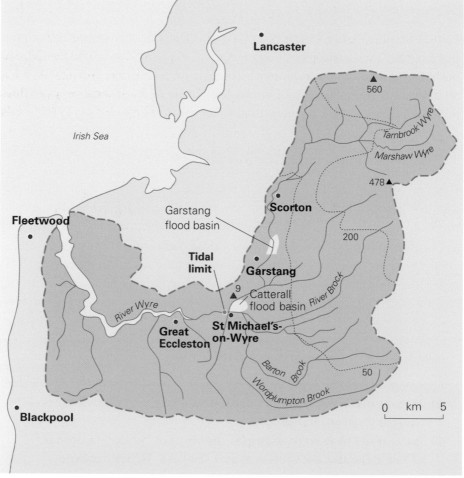

Figure
7.15

Sluice gates on the River Wyre at Garstang

Behind the sluice gates, extensive areas of the valley floor have been reserved as flood basins. The Garstang basin has a capacity of 1.3 million m³ and the Catterall basin can store up to 1.7 million m³. The scheme gives protection against a 1 in 50 year flood event. Figure 7.16 shows the impact of the scheme on peak flows at St Michael's-on-Wyre, a village that was severely affected by the 1980 floods.

Figure
7.16

Impact on discharge at St Michael's on Wyre

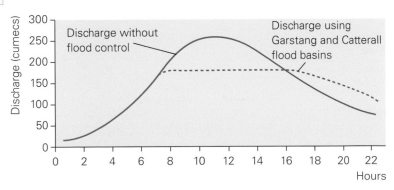

Activity 6

(a) Using the evidence in Figure 7.14, suggest three reasons why St Michael's-on-Wyre has a high risk of flooding.

(b) Study Figure 7.16. Describe and explain the impact of the flood management scheme at Garstang and Catterall on peak flows at St Michael's-on-Wyre.

Problems with structural approaches

The effectiveness of structural approaches to river management has come under critical scrutiny in recent years. Dam construction, channelisation and other hard-engineered structures are extremely costly. In the UK, the Environment Agency currently spends £320 million a year on flood control measures of this type. With climate change and increased flood frequency in the future, continued spending on hard-engineered structures will be unsustainable.

Structural approaches bring other problems too. Invariably they are piecemeal, aimed at modifying small sections of the course of a river, most often the channel and adjacent floodplain. As we have seen, streams and rivers are dynamic systems. The interdependence of their parts means that alterations to the flow pattern or channel characteristics of a river have impacts that create problems elsewhere. These feedback responses are found in all natural systems.

Non-structural approaches

Sustainable solutions to problems of flooding and erosion are more likely to be achieved by working with rivers rather than against them. Non-structural approaches offer an alternative to hard engineering. Streams and rivers are viewed as systems that require management at the drainage basin scale, rather than piecemeal intervention on floodplains and short stretches of channel. This integrated approach is based on geomorphological, hydrological, ecological and economic principles. It meets most of the criticisms of the structural approach and is in tune with the popular notion of working with natural systems, rather than against them.

Non-structural approaches have diverse aims, including:
- flood management (not just flood prevention)
- conservation of wildlife and wildlife habitats
- improving water quality and pollution control

Typical non-structural approaches to river management are:
- control of land use in headwater regions
- land-use zonation on floodplains
- limited structural controls, such as relief channels and flood basins
- flood mitigation through warning systems, insurance protection, risk awareness and publicity campaigns, and household responsibility, e.g. sealing doors and airbricks with sand bags or other protection

Meanwhile, **river restoration policies** aim to dismantle structural controls such as levées and channel realignment that divorce rivers from their floodplains and floodways. This move to 'free' rivers from artificial constraints is motivated partly by environmental reasons (habitat conservation) and partly to reduce the hazards posed by the control structures themselves.

In England and Wales, the Environment Agency is currently developing Catchment Flood Management Plans for all drainage basins. These plans, comparable with existing Shoreline Management Plans for coastal areas, should be developed for entire drainage basins by 2007–08. They will use an integrated approach, relying on a combination of non-structural measures and hard-engineered structures that have been assessed as environmentally and economically sound.

Management of floodplains

Floodplain management means exactly that — management, rather than construction — and is favoured as the most effective means of protecting people and property. At the same time, the aim is to reduce the environmental impacts of hard-engineered structures. Instead of controlling rivers with dams and levées, floodplain management identifies land uses that are compatible with the natural risks. This is done by land-use zoning, building codes, special planning laws for construction, health regulations and erosion control.

Drainage basin management

Managing floods requires large-scale land-use management within drainage basins. Without management, deforestation increases storm runoff and the frequency and size of flood peaks. Uncontrolled road building also increases surface runoff and accelerates soil erosion and gullying. Hard surfaces created by urban development (e.g. roofs, car parks and retail parks) have similar effects, increasing runoff and the risks of flooding. Management of drainage basins involves:

- maintaining or re-establishing the natural vegetative cover
- avoiding road construction and logging
- avoiding development along streams, on unstable slopes, and on highly erosive soils

Management practices also conserve natural water storage reservoirs within catchments. As a result, drainage basins act like sponges, storing water and releasing it slowly. This reduces the rapid runoff that not only causes downstream flooding but also contributes to siltation and habitat destruction.

Drainage basin management: the Cosumnes River, California

The Cosumnes River (Figure 7.17) is the last undammed river flowing out of the Sierra Nevada in California. Until recently, flood management on the river had focused on structural controls. An elaborate levée system separated the river from its floodplain and natural braided reaches had been engineered to form a single narrow channel.

Figure
7.17 **The Cosumnes River basin, California**

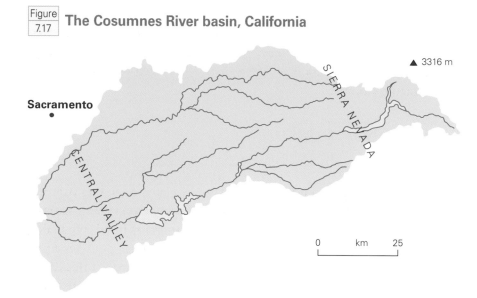

Confining the Cosumnes River within an artificially narrow corridor increased the erosive energy of the river. The result was rapid incision of the river channel, which in places created near-vertical stream banks up to 8 m high. Trees, which formerly stabilised the banks, were undermined and threatened the levées. The artificial channel also removed alluvial bars that were once important spawning areas for salmon and trout.

Overall, the flood control structures on the Cosumnes River were more damaging than protective. It is no coincidence that in 1997 the highest flows ever in the river were recorded. Twenty-four levées were breached and thousands of people had to be evacuated.

A new approach to management

The 1997 floods prompted a re-evaluation of flood-management strategies for the Cosumnes River. Structural controls no longer appeared to work and maintaining the levée system simply led to further channel incision, erosion and bank collapse. Moreover, the levée system required constant upkeep and maintenance.

It was argued that if the Cosumnes River could be reunited with its floodplain, extensive floodplain storage areas would become available, reducing peak flows and flood hazards. At the same time, valuable ecosystems would be recreated on floodplains. Setting back levées would allow the river to spill onto its floodway at high flow and reduce or reverse the damaging channel and bank-erosion problems.

Other measures include flood-proofing homes and businesses and relocating flood-prone housing and commercial activities.

Finally, broadening the floodplain along the Cosumnes River, rather than constricting it with levées, would have other benefits. For example, it would increase groundwater infiltration and provide much needed water resources in south Sacramento County.

Flood mitigation

Flood forecasts and flood warnings

The Environment Agency (EA) is the principal flood defence operating authority in England and Wales. Following severe flooding during Easter 1998, it established a National Flood Warning Centre (NFWC) to improve flood forecasting, flood warning and communications. The flood warning system provides a plan of action for the public to reduce the potential damaging effects of floods.

In order to raise public awareness of flood risks, the EA publishes maps on its website showing floodplain areas, i.e. areas where there is at least a 1% risk of flooding. There is also a telephone Floodline on which the public can find out about current risks and report flooding. The EA has a four-stage system of flood warnings, shown in Table 7.2. Floods in the Wharfe valley, Ilkley are shown in Figure 7.18.

Table 7.2 Flood warning codes

Code	Action
Flood Watch	Flooding is possible. Be aware! Be prepared! Watch out!
Flood Warning	Flooding of homes, businesses and main roads is expected. Act now!
Severe Flood Warning	Severe flooding is expected. Imminent danger to life and property. Act now!
All Clear	The 'all clear' is issued when flood watches or warnings are no longer in force. Floodwater levels are receding. Check it is safe to return. Seek advice.

Floods in the Wharfe valley, Ilkley, West Yorkshire

Activity 7

Log on to the Environment Agency's website at: www.environment-agency.gov.uk
(a) Find out what flood warnings are currently in force by region in England and Wales.
(b) Present the results as a table and comment on them.
(c) Download a copy of a flood hazard map for your locality.
(d) Explain why many floodplains are not defined as hazardous.
(e) Plan a study to investigate local residential areas where flooding is a potential hazard. Design a questionnaire for residents at risk from flooding that investigates:
  - past floods and their impact
  - the perception the residents have of the flood hazard
  - the preparedness of the residents for a flood

Flood insurance

It is estimated that houses on floodplains in the USA have, during the term of a 30-year mortgage, a 26% chance of suffering flood damage. This is a relatively high risk. One mitigating option for people exposed to flood risks is flood insurance.

In 1968, the US Congress, alarmed at the escalating costs of assistance after flood disasters, set up the National Flood Insurance Program (NFIP). The NFIP not only provides flood insurance to households and businesses but also promotes floodplain management and the mapping of flood hazard areas. Any organisation or individual requiring government assistance for building or improving property in flood hazard areas must purchase flood insurance. Typical flood insurance provides cover against damage to property, loss of life, debris removal, relocation and the cost of materials and equipment, such as sand bags and pumps, to combat flooding.

River restoration

In the past, floodplains were viewed as being completely separate from the active channel of a river. Rivers and their **floodways** — the dry zones between levées that convey floodwaters — have been the focus of construction and control. Meanwhile, the fertile, flat and 'reclaimed' floodplain lands have attracted human activities, such as agriculture, commerce and residential development. As river engineering works have become more intrusive and rivers more isolated from their floodplains, there has been a corresponding loss of wetland habitats and a reduction in biodiversity.

River restoration does not have a universal meaning (Table 7.3).

Table 7.3 **Terminology applied to different scales of river restoration**

Term	Definition
Full restoration	The complete reversal of engineered and structural alterations, returning the river to its original state
Rehabilitation	Partial return to the pre-altered state, perhaps only on a river reach scale
Enhancement	Improvements to environmental quality on a small scale, such as individual sites
Mitigation	A small-scale measure that mitigates some of the harmful effects of a current site development
Creation	Development of a feature that did not previously exist at the site and that will improve the conservation value

Full restoration of a river to its natural state is rare. Most rehabilitation projects aim to restore only short stretches of a river. In the UK, two small demonstration projects have been completed: one on the River Skerne at Darlington and the other on the River Cole in Wiltshire.

The River Skerne

The River Skerne flows through the centre of Darlington in northeast England (Figure 7.19). Over the past 200 years, it has undergone straightening and deepening for flood control and drainage. Much of the floodplain was raised above the river by waste tipping and became occupied by housing, gas and sewerage pipes and electricity cables. The habitat diversity, biodiversity and visual appeal of the river were severely degraded.

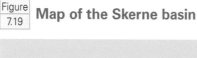

Figure 7.19 **Map of the Skerne basin**

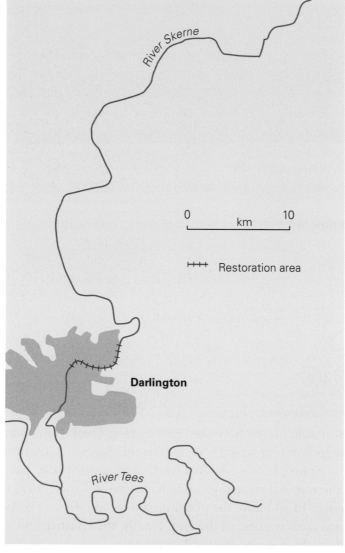

The project aimed to restore a 2 km stretch of the River Skerne, in the town. This meant improving habitat diversity, water quality, landscape and access for the community. The work involved:

- re-meandering the course of the river
- re-profiling the river bank to give more natural shapes
- lowering the floodplain to store water at high flow (Figures 7.20 and 7.21)

| Figuro 7.20 | **River Skerne restoration** |

(a) An unrestored reach of the River Skerne, with reinforced concrete banks and pipework

(b) A restored reach of the River Skerne

Early results of the restoration project have been encouraging. Shallow flooding of planted areas has helped to remove silt from the stream channel. Pool–riffle sequences have been created in the channel and water quality has been improved following the removal of sewage outfalls. In consequence, there is much greater biodiversity. In addition, the planting of trees, shrubs and bulbs has enhanced the visual attractiveness of the area. Finally, the construction of footpaths has opened-up the valley to the local community.

Figure 7.21 **Restoration of the River Skerne, Darlington**

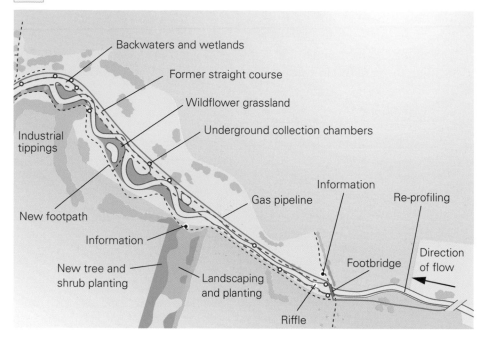

Backwaters and wetlands

Former straight course

Wildflower grassland

Underground collection chambers

Industrial tippings

Information

Re-profiling

Gas pipeline

New footpath

Information

Direction of flow

New tree and shrub planting

Footbridge

Landscaping and planting

Riffle

Activity 8

Complete a table (similar to Table 7.4) to summarise the effects of human activities on stream flow and the physical environment.

Table 7.4 **Effects of human activities on stream flow and the physical environment**

Action	Impact on stream flow	Impact on physical environment
Dam building		
Channel realignment		
Sluice gates and washlands		
Flood-relief channels		
Bank reinforcement		
Levées		
Deforestation		
Urbanisation		
Water abstraction		
River restoration		